From Asphalt to AI:

The Future of Paving Technologies

Verses Kindler Publication

Verses Kindler Publication.

Website: www.verseskindlerpublication.com

From Asphalt to AI:

The Future of Paving Technologies

By: Mr. Pruthvi Yannam

ISBN: 978-93-49532-54-0

1st Edition

Price: INR249/$13

Disclaimer

From Asphalt to AI: The Future of Paving Technologies is authored by Pruthvi Yannam. The entirety of this published work comprises the original content of the author, who has undertaken diligent efforts in its editing. The characters depicted herein may be fictitious or inspired by real events; however, they are not intended to cause offense or represent any prejudice against any caste or system. Any similarity to the names of actual individuals, locations, or institutions is purely coincidental and serves the narrative. The author bears sole responsibility for the originality of this work; the publisher assumes no liability thereof.

From Asphalt to AI: The Future of Paving Technologies

"The future of construction is not just about building structures — it's about building the future of our planet."

- Author

Index

PREFACE

As I stand on the brink of a new construction era, I am excited and compelled to share my thoughts on the profound changes unfurling before us. It is hard to ignore the impact of cutting-edge technologies like artificial intelligence, robotics, the Internet of Things (IoT), and smart materials that are revolutionizing our industry. Each of these innovations is a tool and a catalyst for transformation, reshaping how we conceive, design, and execute projects.

Asphalt is crucial in infrastructure development, providing a durable and flexible material for road surfaces, parking lots, and driveways. Its weather-resistant properties enhance safety by offering better traction in various conditions. Additionally, asphalt recycling promotes sustainability, reducing waste while maintaining quality, making it an essential component for modern transportation systems and urban planning. It has immense potential for use in the construction field. Innovations clubbed with an advanced variety of asphalt can transform this field.

In my years of experience, I have witnessed how innovation can streamline processes, reduce costs, and improve project outcomes. The days are gone when construction was bogged down by inefficiencies and outdated practices. Today, we have the power to harness technology to create workflows that are precise, more efficient, and sustainable. They are also responsive to the needs of the clients and the environment.

Throughout this book, I will share real-world examples and case studies that illustrate the successful integration of these technologies within the construction industry. From smart building systems that enhance energy efficiency to robotics that take on hazardous tasks, these stories are a testament to the brilliance of human ingenuity paired with technological advancement. They demonstrate that the construction of the future is not only possible but is already being realized by forward-thinking professionals.

However, it is essential to recognize that with these advancements comes the responsibility for construction professionals to adapt. As the market continuously evolves, those who fail to invest in innovative solutions risk being left behind. Embracing new technologies is no longer a choice. It is a necessity if we aim to remain competitive and relevant in this rapidly changing landscape.

It's essential to emphasize our responsibility for sustainable development in protecting the environment. Every construction project has a direct influence on environmental health, and we must be mindful of the consequences of our actions. Adopting sustainable practices is vital to ensure a healthier planet for future generations.

In this journey through the pages of this book, I invite you to explore how we can harness the power of innovation to push the boundaries of what is possible in construction. Let's pave the way to a future where creativity and technology converge seamlessly, enabling us to build structures and shape society for the better.

CHAPTER 1

> ***Asphalt paving has significantly evolved from the labor-intensive methods of the past. Previously, the process required extensive manual efforts, including detailed mix designs, handwritten bids, and a reliance on intricate calculations. Workers spent countless hours on quality testing and documentation, often needing to rewrite entire proposals over minor changes. Today, advancements in technology have transformed these practices. The introduction of automated software and AI streamlines mix design and bidding processes, enabling quicker and more efficient project management. This shift has allowed the industry to move away from traditional methods, enhancing productivity and quality, ultimately leading to a more modern approach to asphalt paving.***
>
> ***Frank Surface,***
> ***Chairman, Superior Paving Corp.***

BLUEPRINT FOR TOMORROW: BUILDING WITH INNOVATION AND SUSTAINABILITY

This book is a blueprint for tomorrow and serves as a comprehensive roadmap for the future of construction, where technological advancements and sustainable practices converge to shape a more resilient, efficient, and eco-friendly built environment. This book explores the transformative impact of cutting-edge innovations in construction materials, design, and methods while emphasizing the importance of sustainability and environmental responsibility. As the world faces the challenges of climate change, urbanization, and resource scarcity, the book outlines how the construction industry can adapt and thrive by adopting new technologies, optimizing existing practices, and creating infrastructure that serves both people and the planet.

Each chapter of the book represents a critical layer of construction's evolving foundation, focusing on the transformative elements that shape the future of infrastructure. Here's a brief explanation of how each layer plays a vital role in advancing construction practices.

Materials are the building blocks of progress. Innovation in materials is a fundamental layer that serves as the foundation for all modern construction projects. Through this book, I have explored the evolution from traditional building materials to high-performance and sustainable alternatives. Advances like self-healing concrete, recycled asphalt, and green insulation are reshaping how structures are built. These materials enhance durability and functionality and also minimize environmental impact, paving the way for more eco-friendly and cost-efficient buildings.

Technology revolutionizes construction practices. Technological advancements are transforming the construction industry, and the book covers the integration of Building Information Modeling (BIM), robotics, automation, and 3D printing. These innovations allow for more precise planning, reduced errors, and faster execution of projects. Technology also streamlines communication, improves safety, and enables the creation of complex, sustainable designs that would have been impossible with traditional methods. One of the highlights is how technology enables construction to be smarter, more efficient, and more adaptable.

From self-healing concrete to 3D-printed structures, the reader can explore how innovative materials are revolutionizing the way we build. Readers will also learn about the integration of nanotechnology, biomimicry, and smart fabrics that respond to environmental factors and reduce the need for maintenance.

The role of robotics and drones in construction sites for improving efficiency, safety, and precision needs to be understood in detail as each day comes with a new revelation. I tried to help the reader comprehend how robotic bricklaying, automated heavy machinery, and 3D printing are streamlining workflows and lowering costs.

On diving deep into BIM technology, one realises that it is a game-changer for design, construction, and maintenance. The section on BIM highlights how it improves collaboration, reduces errors, and allows for precise planning and resource allocation from project conception to completion.

Sustainability is the building for the future. Sustainability is a crucial layer that focuses on minimizing environmental impact while ensuring that buildings and infrastructure are resilient, energy-efficient, and resource-conscious. The chapter explores the shift toward green building practices, the use of renewable energy systems, and waste reduction techniques like recycling materials and using low-carbon footprint solutions. Sustainability not only addresses climate change but also creates long-term value by reducing operational costs and improving the livability of spaces.

The reader will learn how architectural designs that optimize natural lighting, insulation, and ventilation can drastically reduce energy consumption. Discover examples of green building certifications such as LEED and how they push the industry toward sustainability.

The growing trend of incorporating solar panels, wind energy systems, and geothermal heating and cooling into buildings has been examined in detail. Understand how buildings can be self-sustaining and even contribute surplus energy to the grid.

Take another deep dive into the importance of using recycled materials in construction, from reclaimed wood and recycled steel to repurposed glass. Explore how construction waste can be minimized and repurposed through innovative systems like closed-loop recycling.

As construction evolves, the workforce layer plays a pivotal role in adapting to new tools, techniques, and technologies. I have emphasized the importance of upskilling the workforce to handle advanced machinery, digital tools, and innovative materials. It also discusses the need for inclusive leadership and diversity in construction to foster a collaborative and adaptable environment. A skilled workforce ensures the smooth integration of new technologies and the successful execution of complex projects.

Safety and risk management protect people and projects. The safety and risk management layer is essential to the success of any construction project. This part explores the evolution of safety standards, smart safety gear, and data-driven risk management systems. By embracing predictive analytics and real-time monitoring technologies, construction leaders can anticipate potential hazards, prevent accidents, and ensure a safer working environment for everyone involved.

All construction processes require efficiency and precision. The construction process is the operational layer that governs how materials, labor, and technology create structures. The reader can examine lean construction principles, modular construction, and prefabrication techniques that increase speed and accuracy while reducing waste. Enhanced project management practices, such as agile methodologies and real-time monitoring, ensure that projects stay on track, within budget, and adhere to quality standards.

The focus has also been on how urban developments are evolving to be more resource-efficient, biodiverse, and climate-resilient. It covers how cities are integrating green spaces, eco-friendly transportation networks, and waste management systems to reduce environmental impact.

Explore how construction practices are adapting to the increasing frequency of extreme weather events, rising sea levels, and more intense heat waves. Understand how flood-resistant foundations, heat-reflective surfaces, and climate-adaptive materials are becoming essential for future-proof infrastructure.

Learn about green roofs and permeable pavement that reduce runoff and contribute to sustainable water management.

As automation changes the landscape of construction, how will workers be impacted? What is the role of leaders in the field of construction?

A major section explores the importance of upskilling the workforce, combining human ingenuity with machine precision for more effective construction projects. The book will also emphasize the need for inclusivity in both the design of structures and the construction workforce looking at the future of accessible buildings for people of all abilities, ensuring that technological advancements also serve diverse populations.

Featuring real-world examples of sustainable skyscrapers, eco-cities, and green infrastructure projects, the book highlights visionary construction efforts already underway around the globe. These case studies offer valuable insights into how the latest innovations are being put to work and shaping the cities of tomorrow.

Understanding the financial impact of green construction practices, I have demonstrated how sustainable buildings are not only good for the environment but can also generate long-term savings through reduced energy costs, increased durability, and higher property values. The leaders play a crucial role here.

A look at how governments are shaping the future of construction through building codes, sustainability mandates, and incentives for environmentally friendly building practices is also an important factor in the construction field. Thus this book also provides an overview of what leadership in asphalt road construction entails.

Through this book, I also aim to inspire construction professionals, innovators, and students alike to embrace sustainable practices, explore new technological solutions, and shape a future where the built environment works in harmony with the planet's resources.

Client and stakeholder engagement is the foundation for collaboration for success. The client and stakeholder engagement layer focuses on how effective communication and collaboration lead to better project outcomes. The book explores how stakeholder involvement, community engagement, and transparent communication can shape the direction of a construction project. It highlights the importance of aligning project goals with the needs of clients, communities, and regulatory bodies to create lasting, positive impacts.

Resilience and adaptability help in designing for the unpredictable. Resilience and adaptability are the layers that ensure long-term success. It covers how climate resilience, adaptive reuse of structures, and future-proofing designs are becoming critical in an increasingly uncertain world. It discusses how construction is adapting to extreme weather events, resource shortages, and technological shifts, ensuring that infrastructure can withstand both environmental changes and evolving societal needs.

Together, these layers form a comprehensive foundation for the future of construction, highlighting the importance of innovation, sustainability, and forward-thinking approaches in creating a built environment that is both functional and resilient for future generations. Each chapter represents a unique aspect of construction's evolution, working in tandem to transform the way we design, build, and live.

Finally, it is important to understand that leadership in asphalt road construction is crucial for ensuring that projects are completed successfully, on time, within budget, and with a focus on safety, sustainability, and innovation. Effective leadership can drive excellence in both the technical and managerial aspects of asphalt road construction, guiding teams, maintaining high standards, and addressing the challenges that arise in such large-scale projects.

Visionary leadership is essential in shaping the future of infrastructure. A strong leader in asphalt road construction must have a clear vision for the future of the industry. This includes staying ahead of technological advancements like smart materials, automated paving, and sustainable construction practices. Leaders must not only manage current projects but also plan for the integration of new technologies that can enhance efficiency, durability, and environmental sustainability.

Sustainability focus is crucial in today's times. As environmental concerns become more pressing, leaders in this field must prioritize sustainable construction practices. This includes adopting recycled asphalt materials (RAP), using warm-mix asphalt to reduce emissions, and implementing green paving techniques that minimize ecological impact.

Technical expertise is an essential prerequisite for innovation. In-depth knowledge of materials and techniques is required. A leader must have a solid understanding of the asphalt mix design, paving machinery, and construction methods. This knowledge ensures that projects meet the required performance standards for durability, safety, and smoothness.

Leaders in asphalt road construction need to embrace new technologies that improve the efficiency of production and quality control. For instance, adopting geospatial technologies for real-time tracking and measurement, or automated paving equipment, can help reduce errors and improve the precision of road construction.

With the rise of recycled asphalt pavement (RAP) and warm-mix asphalt, forward-thinking leaders are adopting these methods to reduce costs and environmental impact. They should also explore biodegradable materials and self-healing asphalt to enhance the longevity and sustainability of roads.

Creating a safety culture is a prerequisite for any project. Leaders in asphalt road construction must prioritize safety at all stages of the project. Ensuring compliance with OSHA standards, implementing site-specific safety plans, and providing proper training to workers are critical to minimize accidents and injuries.

Construction projects often face unpredictable challenges such as inclement weather, supply chain disruptions, or design changes. A strong leader needs to anticipate potential risks and have contingency plans in place to address issues before they escalate, keeping the project on track and reducing costly delays.

Leadership in asphalt road construction is not just about technical knowledge but also about building and managing effective teams. A good leader must foster collaboration between engineers, designers, workers, suppliers, and local governments to ensure seamless project execution.

Ensuring clear communication with stakeholders is vital for successful project completion. This includes interacting with clients, regulatory agencies, local communities, and contractors. Effective communication ensures that everyone is aligned with the project's goals, timelines, and expectations.

Leaders must oversee the quality control (QC) processes throughout the project. This involves conducting regular inspections, using advanced testing methods for asphalt mix quality, and ensuring the final surface meets all required standards for smoothness, compaction, and durability.

Leadership includes being hands-on with the testing of asphalt samples during the construction process, making sure that asphalt properties like density, viscosity, and resilience are monitored. Testing ensures the road performs under the weight of traffic and various weather conditions.

A strong leader must ensure the project is delivered within budget. This involves negotiating with suppliers for cost-effective asphalt mixes, managing labor resources, and ensuring efficient use of machinery and equipment. Leaders need to balance quality and cost without compromising safety or performance.

Beyond the initial construction phase, leadership involves understanding the lifecycle costs of asphalt roads. This includes planning for future maintenance and rehabilitation, ensuring that long-term durability is built into the design and materials used.

Asphalt road construction leaders must be committed to reducing the environmental impact of their projects. This means using recycled asphalt (RAP) to reduce material waste, exploring the use of warm-mix asphalt to lower energy consumption during production, implementing dust control systems and ensuring that construction activities do not harm surrounding ecosystems is a must.

Leaders should advocate for porous asphalt or permeable paving, which allows water to pass through the surface, helping to manage stormwater runoff and reduce flooding.

In asphalt road construction, leaders must be responsive to the needs of the community and stakeholders. This may include conducting public meetings or open houses to inform the public about project impacts, timelines, and benefits.

As many asphalt road construction projects occur in heavily populated areas, leaders must minimize disruption to daily life. This includes managing traffic reroutes, keeping construction noise to a minimum, and maintaining local businesses' access to traffic.

A leader who is transparent with all parties—whether clients, workers, or the public—builds trust and enhances the reputation of the company or organization they represent.

A key part of leadership is ensuring that asphalt road construction projects are completed on schedule and to the agreed-upon quality and standards. This builds the confidence of stakeholders, leading to repeat business and positive referrals.

The asphalt industry is evolving, with increasing attention to sustainability, efficiency, and smart technology. A strong leader must be adaptable, keeping up with changing regulations, market trends, and technological advancements to remain competitive and innovative.

Leadership in asphalt road construction goes beyond just managing workers and overseeing day-to-day operations. It requires a blend of technical expertise, strategic thinking, sustainability consciousness, and an ability to foster collaboration and effective communication.

Leaders who combine these qualities can drive the industry forward, creating safer, more durable, and environmentally responsible roadways that serve the needs of communities and the economy for years to come.

The book concludes by offering a vision for the future of construction, where technology, sustainability, and collaboration create an industry capable of addressing the needs of a growing global population while minimizing its environmental impact. The book challenges readers to consider their role in this transformation — whether as builders, architects, policymakers, or citizens — and how they can contribute to a construction industry that is both forward-thinking and environmentally responsible.

CHAPTER 2

THE DIGITAL TWIN OF THE ROAD: 3D MODELING AND VIRTUAL CONSTRUCTION.

A digital twin is a virtual representation of a physical object, system, or process that adapts using real-time data, allowing for functions such as simulation, monitoring, and optimization. These virtual models accurately reflect real-world entities, including products, buildings, or entire manufacturing facilities, through continuous updates from sensors, IoT devices, and other sources. Digital twins enable simulations, predictive analytics, and "what-if" scenarios, thus improving decision-making and optimization.

Digital twins are making a significant impact on the real world, ranging from the personalized level of an individual digital patient to broader applications that enable entire nations to prepare for climate change. This innovative technology is being embraced by companies and organizations across various sectors, including energy, hospitality, government, and healthcare, as they find creative methods to enhance efficiency and meaning in our work and lives. Below are four real-world examples of digital twin technology currently in use.

3D modeling and digital twin technology have become vital in the planning, design, and construction of asphalt pavements. These advanced tools enable more efficient workflows, improved accuracy, and reduced costs throughout the pavement lifecycle, including the planning, design, construction, and post-construction phases.

In the initial planning phase, 3D modeling allows engineers and project managers to visualize the entire project from different angles before any physical work begins. Engineers can create detailed 3D construction site models using aerial imagery, LiDAR scans, or existing 2D blueprints. This helps in understanding the topography, drainage patterns, and existing infrastructure. With 3D models, planners can simulate various scenarios to assess environmental impacts and comply with regulations. They can analyze how changes in design affect stormwater runoff and ecological balance. Stakeholders, including local governments and community members, can better understand the project through 3D visualizations. This transparency can improve communication and receive necessary approvals more efficiently.

During the design phase, digital twin technology plays a pivotal role in ensuring precision and efficiency. A digital twin is a virtual representation of a physical asset that simulates its behavior in real time. For asphalt pavements, digital twins can simulate various material properties and performance over time under different load conditions and weather scenarios. For material selection, engineers can assess various asphalt mixes and structural designs using 3D modeling software. This enables them to optimize the design for durability, load-bearing capacity, and sustainability. Digital twins allow for the analysis of the pavement's lifecycle, predicting wear and maintenance needs over time. This insight enables engineers to create more durable designs tailored to anticipated traffic patterns.

In the construction phase, 3D modeling and digital twins enhance project execution by improving efficiency and quality control. With detailed 3D models, construction teams can understand exact dimensions and specifications, reducing materials waste and ensuring accurate asphalt placement resulting in precision in construction. Digital twin technology can be employed to monitor construction activities in real time. Sensors embedded in construction equipment can provide data on pavement temperature, moisture levels, and compaction, ensuring that the asphalt is laid down perfectly.

As the pavement is being constructed, the digital twin can simulate the expected performance, allowing engineers to compare real data against predictions. This helps quickly identify any issues that arise during the construction process, enabling timely interventions to ensure quality maintenance and assurance.

Once the asphalt pavement is in place, the benefits of 3D modeling and digital twins continue which is the post-construction phase. Digital twins can continually gather data about pavement condition, facilitating predictive maintenance strategies. By analyzing this data, authorities can schedule repairs before significant issues occur, extending the life cycle of the pavement. The ongoing use of digital twins can help monitor how the pavement performs under different traffic and environmental conditions. This data is invaluable for future projects, providing insights into what design elements work best.

Thus, 3D modeling and digital twin technology have revolutionized the approach to planning, designing, and constructing asphalt pavements. The integration of these technologies facilitates better visualization, enhances decision-making, reduces risks, and improves overall project efficiency. As technology advances, these tools will likely become even more integral to pavement engineering, leading to smarter, more sustainable infrastructure development.

Virtual construction, often associated with Building Information Modeling (BIM) which I shall be talking about in detail, and other digital technologies, offers numerous benefits that significantly enhance accuracy, efficiency, and cost-effectiveness in the construction industry. Virtual construction allows for the creation of detailed 3D models of a project before construction begins. This enables stakeholders to visualize the final product, reducing the likelihood of design errors. With advanced software, teams can identify and resolve potential conflicts (clashes) between different building systems (e.g., electrical, plumbing, HVAC) early in the design phase. The integration of real-time data into the virtual model ensures that any changes to the design are immediately reflected, minimizing discrepancies between the model and the actual construction.

Virtual construction promotes better communication among architects, engineers, contractors, and clients. Everyone can access the same model, making collaboration smoother and more effective. The use of virtual tools helps streamline workflows. Task assignments can be structured around the model, allowing for a more efficient allocation of resources and scheduling. Virtual construction enables teams to simulate construction sequences and processes, leading to optimized scheduling and more effective logistics management.

By catching design errors and potential conflicts early, virtual construction significantly reduces the need for rework, which can be one of the largest sources of costs in construction projects. The detailed models can be used for accurate quantity takeoffs, enabling more precise cost estimates and aiding in budget management throughout the project lifecycle. Virtual construction allows for better planning of materials, labor, and equipment, which leads to minimized waste and reduced costs.

Virtual models can be used to visualize safety risks and develop better safety plans, allowing teams to identify and mitigate hazards before construction starts. Virtual environments can be used for training personnel, helping workers understand potential dangers and safe practices in a realistic but controlled setting. Virtual construction enables simulations that can analyze energy consumption and material use. This facilitates the design of more sustainable buildings and helps in adhering to environmental regulations.

By leveraging virtual construction technologies, stakeholders in the construction industry can ensure that projects are executed more smoothly, efficiently, and economically, ultimately leading to higher-quality outcomes.

As reported by Deloitte, IDC anticipates that by 2022, 40% of IoT platform vendors will incorporate simulation platforms, systems, and capabilities to develop digital twins. Additionally, 70% of manufacturers are expected to leverage this technology for process simulations and scenario analysis. If your company is currently utilizing IoT technology or considering its adoption, it's essential to monitor how digital twins may drive transformation in your industry going forward.

Singapore, as an island nation, faces significant challenges from rising sea levels. To tackle these issues, the country has implemented an advanced integrated digital twin infrastructure. This infrastructure features a precise and reliable terrain model that aids the national water agency in managing resources, planning, and coastal protection. In a pioneering effort to utilize the digital twin concept, Singapore has developed a comprehensive virtual representation of the country, encompassing building and transportation models. This innovative approach enables one of the world's top three smart cities to devise strategies for tackling climate change, particularly in response to the threats posed by rising sea levels

Faced with rampant supply chain delays, labor shortages, and inflated material costs, the stakes for builders are at an all-time high. Bad data and poor decision-making can lead to expensive delays and rework. Digital twin and AR technology allow the construction industry to optimize project data, streamline collaboration, and better visualize projects from design through to operations and maintenance. By using AR to bring valuable BIM data to the field, contractors are able to capture and communicate design errors in just a few clicks, allowing stakeholders to resolve issues quickly and avoid costly rework.

Visualizing designs in real-time can be done by using any mobile device to overlay building information modeling (BIM) data in the field using AR to quickly inspect design and installed conditions and stop rework in its tracks. Streamlined virtual design and construction (VDC) workflows are extremely helpful. The issues are captured in the field marked directly within the model, reducing downtime and making office-to-field communication seamless. Immersive safety training transforms job-critical learning and training into interactive, immersive environments across key scenarios. Maintain critical model and project data that will improve digital twin operation and maintenance long after completion. DPR Construction leverages AR to empower field teams. DPR, an ENR Top 10 Contractor, is integrating AR and immersive tech into the project lifecycle to bring valuable BIM data to the field in real-time to improve team performance and reduce rework.

Digital twin technology helps builders, planners, and operators across cities worldwide better understand and optimize these spaces for public use. By using advanced, interactive models and live IoT data, stakeholders are able to simulate traffic flow, mobility patterns, and even the effects of climate change and shifting landscapes surrounding key infrastructure like airports, roads, and transportation hubs. From individual facilities to smart cities, digital twins are helping owners, operators, and policy-makers manage large volumes of valuable data that will allow them to better equip our infrastructure for future demands.

Digital twins integrate geospatial, modeling, and sensor data to provide real-time visualizations of spaces, equipping decision-makers with essential information to enhance operations and allocate resources effectively. By utilizing digital twins, planners and designers can simulate the movement of people, vehicles, and other elements within a space or even across an entire city. This capability allows them to predict how these movements might change in response to alterations in the environment, thereby guiding key decisions related to design and infrastructure upkeep. When combined with data from IoT sensors and energy usage, this modeling helps teams identify inefficiencies, enabling them to minimize emissions, waste, and water consumption.

According to ABI Research, it is expected that over 500 cities will implement digital twins by 2025. Discover how leading players in the smart city sector are using Unity to realize urban digital twins.

The use of Augmented Reality (AR) in the asphalt paving industry presents a transformative opportunity to enhance various aspects of the paving process. Here's a detailed exploration of how AR can be applied, its benefits, and its potential challenges.

AR can provide real-time visualizations of pavement designs overlaid onto the existing environment. Contractors and clients can see how the finished project will look in its context. This can aid in decision-making and adjustments before actual work begins. Engineers and planners can create 3D models of proposed asphalt pavements and superimpose them on the construction site. This facilitates better understanding and communication of complex designs.

AR can display real-time data directly on the work site, including temperature readings of asphalt, moisture content, and other critical paving parameters. Workers can make on-the-spot adjustments based on the conditions observed through AR devices. Project managers can use AR to track the progress of paving operations against the timeline and deliverables. Overlays of schedules and progress metrics can help keep the project on track. AR can provide immersive training environments where workers can learn about equipment operation, safety protocols, and best practices without the risks associated with real machinery and live conditions. AR can superimpose safety warnings and instructions directly onto the worker's field of vision, helping to prevent accidents and ensuring safety equipment is used correctly.

Inspectors can use AR to evaluate the condition of asphalt surfaces, identifying cracks, ruts, and wear. Information can be overlaid onto the visual feed of the asphalt, making it easier to pinpoint areas needing repair. AR applications can integrate historical data about the pavement, helping inspectors understand past maintenance and performance issues when evaluating current conditions.

There are innumerable benefits of AR in the Asphalt Paving Industry. By providing immediate access to information and visualizations, AR can streamline operations, reducing rework and improving overall productivity. AR tools improve precision in paving, which can lead to better quality outputs and fewer mistakes that require costly fixes. Visual tools enhance communication among teams, stakeholders, and clients, leading to better collaboration and fewer misunderstandings. By reducing errors and improving efficiency, AR can lead to significant cost savings across projects.

There are many challenges in this area and some considerations are essential to make. Implementing AR requires the workforce to adapt to new technologies, which may involve training and a learning curve. The investment in AR hardware (like smart glasses or devices) and software development can be significant. For AR to be effective, it must be integrated with other technologies (such as GIS and project management systems). This integration can be complex and require substantial technical support. Environmental factors such as lighting and weather can affect the effectiveness of AR applications, potentially limiting their use on-site.

The integration of Augmented Reality in the asphalt paving industry holds considerable promise for enhancing productivity, quality, and safety. While challenges exist, the potential for increased efficiency and improved outcomes could lead to a broader adoption of AR technologies in paving processes. As the industry continues to evolve, leveraging these advanced tools could pave the way for more innovative construction practices.

BIM, Building Information Modeling, is a process that uses a digital representation of a building or infrastructure to facilitate design, construction, and management throughout its lifecycle. It's a collaborative approach involving various stakeholders to create a comprehensive 3D model and manage information for informed decision-making. In the real-world context, while it is widespread, fully integrated BIM for all asphalt paving is still developing, and many departments of transportation, and large construction firms, are moving toward it. The use of BIM to model underground utilities before paving is becoming more common, to prevent costly conflicts during construction.

Here's a more detailed explanation of the key applications of BIM in Asphalt Paving Projects. It is being used for improved planning and design, enhanced project coordination, accurate quantity takeoffs and cost estimation, pavement management and maintenance, and, most importantly, sustainability assessments.

BIM allows for the creation of detailed 3D models of roadways, enabling engineers to visualize and analyze designs before construction. This helps identify potential conflicts and optimize designs for efficiency. This is especially valuable for complex projects involving utilities, drainage systems, and varying terrain. BIM facilitates collaboration among project stakeholders, including engineers, contractors, and owners. By sharing a central model, teams can improve communication and coordination, reducing errors and delays. BIM models can be used to generate accurate quantity takeoffs for asphalt and other materials, enabling more precise cost estimations. This helps control project budgets and reduce waste. BIM can be used to store and manage data related to pavement conditions, maintenance history, and performance. This information can be used to develop effective maintenance plans and optimize pavement life cycles. Integrating BIM with pavement management systems allows for a more data-driven approach to road maintenance. BIM can be used to analyze the environmental impact of asphalt paving projects, including material usage, energy consumption, and emissions. This helps promote sustainable construction practices.There is increased research into integrating Life Cycle Assessments (LCA) into BIM workflows for road pavements, to better understand and minimize environmental impacts. WW Clyde Construction, is a company that is using BIM in horizontal construction, and seeing the benefits of it. This includes the improvement of project planning and sequencing.

Interoperability among various Building Information Modeling (BIM) software platforms presents significant challenges for seamless collaboration in the construction industry. Each platform may use different data formats, making it difficult to share information effectively. This highlights the need for standardized data formats and workflows, particularly in horizontal construction, where alignment across tools is crucial. Additionally, there remains notable resistance to BIM adoption within the road construction sector. Factors contributing to this hesitance include a lack of awareness, training, and perceived complexity, which can hinder progress toward more efficient and modern construction practices. Addressing these challenges is essential for industry advancement.

Despite the challenges, BIM is increasingly being used in asphalt paving projects to improve planning, coordination, and efficiency. While challenges remain, the benefits of BIM are driving its adoption in the road construction industry.

The digital twin technology in road construction in developing countries has a huge scope. Digital twin technology, particularly in the context of road construction and infrastructure management, represents a transformative leap in how developing countries can approach their infrastructure challenges. A digital twin is a dynamic, virtual representation of a physical asset, such as a road or bridge, that is continuously updated with real-time data and can be used for design, construction, maintenance, and monitoring purposes. In developing countries, the adoption of digital twin technology can be a game-changer, helping to address a myriad of infrastructure challenges, including inadequate planning, poor quality construction, frequent maintenance issues, and inefficient resource use. While these countries may face challenges in terms of technology adoption, the scope of digital twins in road construction offers numerous opportunities for growth, sustainability, and enhanced quality of life.

Developing countries often struggle with poor infrastructure, including underdeveloped road networks, bridges, and transportation systems. According to the World Bank, poor infrastructure can hinder economic development by increasing transportation costs, limiting access to markets, and reducing productivity. In many regions, inadequate road networks exacerbate challenges related to poverty, access to education, healthcare, and employment. Moreover, roads in these areas frequently suffer from neglect, with little preventive maintenance, resulting in faster degradation and higher costs for repairs.

Traditional approaches to road construction and maintenance often fall short in terms of efficiency and sustainability. Limited access to skilled labor, materials, and modern tools, combined with a lack of data-driven decision-making, leads to suboptimal road quality and high maintenance costs. These challenges present a compelling opportunity for digital twin technology to provide a more efficient, sustainable, training and capacity building, and cost-effective solution.

Digital twin technology has the potential to revolutionize road construction in developing countries by offering a data-driven approach that improves design, construction, and maintenance practices.

Digital twins can be utilized to address infrastructure gaps in these regions in ensuring improved planning and design, cost savings and resource optimization, proactive maintenance and lifecycle management, disaster resilience and climate adaptation, and enhanced collaboration and transparency.

Road construction in developing countries is often hindered by poor planning and lack of comprehensive data, leading to inefficient designs that do not account for local environmental, social, or economic factors. Digital twin technology can bridge this gap by enabling more accurate and precise planning.

Using 3D modeling and simulation tools, digital twins provide an accurate representation of existing road conditions, including topography, terrain, and infrastructure. These models can incorporate a wide range of data such as soil composition, traffic patterns, weather conditions, and even social factors like population density or future growth projections. With such detailed insights, planners can design roads that are more resilient to local conditions, such as frequent flooding or harsh weather, and more effectively cater to the needs of the local population.

Moreover, the ability to test various scenarios in a virtual environment allows engineers to optimize road designs for factors such as cost, traffic flow, and environmental impact, reducing the likelihood of costly design errors and rework during the construction phase.

Developing countries are often constrained by limited budgets, which makes it essential to maximize the impact of every dollar spent on infrastructure. Digital twin technology offers a solution by optimizing resource allocation and improving project management. By simulating the entire construction process in a virtual environment, digital twins allow stakeholders to anticipate challenges and identify potential inefficiencies early in the project lifecycle. This enables better planning and ensures that resources such as materials, labor, and equipment are used more efficiently, reducing waste and cost overruns.

Furthermore, with accurate digital models of roads, construction teams can avoid mistakes during the building phase, such as incorrect measurements or inefficient material usage. This not only saves money but also ensures that the construction process is completed on time, preventing delays that could increase costs or disrupt the local economy.

In many developing countries, roads suffer from a lack of regular maintenance due to inadequate funding, a shortage of technical expertise, and a lack of data to guide maintenance efforts. As a result, roads quickly degrade, leading to more expensive repairs and a higher frequency of road closures. Digital twin technology offers a solution by providing real-time monitoring of road conditions.

By embedding sensors in the roads and linking them to a digital twin model, infrastructure managers can monitor key indicators such as pavement condition, traffic load, weather, and wear and tear. This real-time data allows for proactive maintenance, as it can trigger alerts when a section of the road is beginning to show signs of stress or damage. This early intervention can prevent more significant problems from developing, extending the lifespan of the road and reducing the need for costly emergency repairs.

In addition, digital twins can help authorities prioritize maintenance efforts based on the most urgent needs. With limited resources available, governments can use data-driven insights to direct funds to the areas where they will have the greatest impact, thus improving the overall efficiency of infrastructure management.

Many developing countries are highly vulnerable to natural disasters such as floods, earthquakes, and landslides, which can cause significant damage to road networks. Digital twin technology can play a critical role in building more resilient infrastructure by simulating how roads and bridges will perform under different disaster scenarios.

For example, in flood-prone areas, digital twins can model how roads will be impacted by rising water levels, enabling planners to design drainage systems that minimize flood damage. Similarly, roads in earthquake-prone regions can be designed and strengthened based on data from digital twin models that simulate seismic activity. This approach helps ensure that roads are more resilient and can withstand extreme weather or natural disasters, reducing the cost of rebuilding and improving the safety of communities.

In developing countries, corruption and lack of transparency can undermine infrastructure projects. The implementation of digital twins can foster greater collaboration and accountability throughout the construction process. Digital twin technology provides a shared, real-time view of the project that can be accessed by all stakeholders, including government agencies, contractors, engineers, and the public.

This transparency allows for better coordination between teams and ensures that the project is progressing according to schedule and budget. Additionally, by providing detailed and accurate documentation of the construction process, digital twins can help reduce corruption by minimizing opportunities for fraud and mismanagement.

One of the significant challenges in developing countries is the lack of skilled workers in the construction and engineering sectors. Digital twins can assist in addressing this issue by providing a platform for training and capacity building. Engineers and construction workers can interact with virtual models of road projects, gaining valuable experience in a safe and controlled environment before working on real projects. This hands-on approach can help improve skills, increase efficiency, and reduce the likelihood of errors during construction.

While the potential of digital twin technology in road construction is vast, its implementation in developing countries will require overcoming several barriers. Access to technology, data availability and quality, and financial constraints can be impediments. Many developing countries face challenges in accessing the necessary technology to create and maintain digital twins, including advanced hardware, software, and skilled personnel. Governments and international organizations can play a crucial role by investing in the necessary infrastructure and providing technical assistance and training. The creation of accurate digital twins depends on the availability of high-quality data. In many developing countries, data collection is often fragmented or lacking, particularly in rural or remote areas. To overcome this, countries must invest in data collection systems, such as sensors and drones that can gather real-time information on road conditions and other relevant factors. The upfront costs of adopting digital twin technology may be a significant hurdle for many developing countries, with limited budgets for infrastructure projects. However, the long-term cost savings and benefits, such as reduced maintenance costs and improved project efficiency, can make digital twins a worthwhile investment in the long run.

Nevertheless, the scope of digital twin technology in road construction for developing countries is enormous. It provides a powerful tool to improve planning, reduce costs, enhance maintenance, and build resilient infrastructure that can withstand the challenges of a rapidly changing world. Despite the barriers to adoption, such as technology access and financial constraints, the potential benefits far outweigh the challenges. By embracing digital twins, developing countries can take significant steps toward improving their road networks, driving economic development, and enhancing the quality of life for their citizens. With the right investments in technology, data, and skills development, digital twins can help these nations build smarter, more sustainable infrastructure for the future.

Digital twin technology is the future.Although a digital twin sounds like the avatar you'd choose for a video game or a virtual reality experience, it's an increasingly mainstream technology that will profoundly shape the world (and beyond) in ways we're just beginning to understand. It's already helping energy companies make their wind turbines more efficient, allowing entire countries to plan for climate change, and even letting fast-food restaurants get that burger out to you just a little quicker.

CHAPTER 3

"In my opinion, I believe that tech will help to solve issues like bidding easier and possibly more accurately by using things like maps, AI software. I think that road replacement might be measured by some device or devices hooked to some sort of self-driving car that helps determine which roads need done again as well. I could see self-driving trucks becoming a very prevalent part of the systems as well. Mostly, within 10 years will simply be a deeper dive into data that we are accumulating from all of these new systems like cameras, fleet maintenance, etc. Those systems will be leveraged into AI systems to determine equipment life spans, etc. (I know that was word salad but it hit one about 4 different things there)"

- *Mike Brewer, Director of IT*

Superior Paving Corp.

THE BUILDER'S PLAYBOOK: INNOVATING FOR THE FUTURE

The construction industry, while critical to the global economy, faces numerous challenges ranging from cost overruns to delays, safety concerns, and the need for sustainable practices. To tackle these issues and stay ahead in an increasingly competitive market, professionals in the construction industry must embrace innovative strategies, technologies, and approaches. This playbook is a guide designed to offer practical strategies and cutting-edge innovations that construction professionals can implement for efficient, effective, and sustainable project delivery.

One basic step in embracing digital transformation is using Building Information Modeling (BIM) which I have already introduced in the previous chapter.

BIM is revolutionizing the construction industry by enabling collaborative project planning, design, and execution. With BIM, professionals can create digital representations of the physical and functional characteristics of a building. It improves accuracy in design, construction, and operations. Building Information Modeling (BIM) has become one of the most transformative technologies in construction. For construction industry leaders, adopting BIM streamlines project planning and design, improves collaboration, and reduces costly errors. To successfully implement BIM ensure that all stakeholders are trained in BIM software, from architects to subcontractors. Invest in interoperable BIM platforms to ensure compatibility across different disciplines and stages of the project. Encourage real-time collaboration between project teams using BIM to reduce delays and rework.

One of the most successful examples of BIM adoption is the Hudson Yards project in New York City. Using BIM, project teams were able to plan and execute complex designs efficiently, cutting down time spent on design changes by more than 50%. The project's use of BIM helped them identify potential issues before they became real problems, minimizing costly delays and rework. The successful integration of BIM in Hudson Yards became a model for other large-scale developments, demonstrating that digital transformation can lead to significant cost savings and improved outcomes.

Here are some useful tactics for leaders. Establish a BIM champion within your organization who can drive adoption and serve as an internal resource. Implement BIM in both pre-construction and construction phases, utilizing it for everything from design visualization to project monitoring and post-construction maintenance. Develop standard operating procedures (SOPs) for integrating BIM into your workflows to streamline its adoption across teams.

An actionable strategy for BIM implementation is a must in any construction project. Invest in BIM software and training to ensure the team can utilize its full potential. Integrate BIM into every phase of the project—from design to construction to maintenance. Use BIM to coordinate between different disciplines (e.g., architecture, engineering, and construction), improving communication and reducing errors.

The use of BIM has expanded with the integration of Augmented Reality (AR) and Virtual Reality (VR) to allow real-time visualization of construction projects. For example, using AR, workers on-site can overlay 3D models of the building over the physical structure, detecting issues before they arise.

Next, we must talk about how modular and prefabricated construction help streamline the build. Modular construction, which involves assembling components of a building off-site and then transporting them to the construction site, offers a wealth of benefits such as speed, quality control, and reduced waste. Prefabrication takes this idea further by creating standardized components that can be assembled in different configurations. Partner with manufacturers of modular components early in the design phase to select the right modules.

Modular and prefabricated construction controls speed and quality. Modular and prefabricated construction methods involve manufacturing building components off-site in a controlled environment, which reduces construction time and improves quality control. To effectively adopt modular construction start with smaller projects to gain experience before scaling up to large, complex developments. Partner with established manufacturers specializing in modular construction to ensure high-quality components are delivered on time. Plan the logistics carefully for transportation, site preparation, and assembly, as these stages can significantly impact the project timeline.

The University of California, Berkeley used modular construction for their Undergraduate Dormitory Project, reducing the overall construction timeline by 50%. With the prefabricated dormitory units built off-site and delivered ready for assembly, the project was completed on budget and ahead of schedule. This approach saved time and ensured a higher level of quality control, as each module was built in a factory setting with more precise standards than could be achieved on-site.

Analyze the project timeline to determine if prefabrication can significantly reduce construction time. Plan for transportation logistics to ensure timely delivery and minimize potential delays. I advise building relationships with modular component suppliers and ensuring they meet your project's specific needs for quality, durability, and timeline.

Also, analyze your project portfolio and identify where modular construction can be applied, especially for repetitive or standardized elements like residential units or office buildings. Incorporate modular construction into your early planning and design phases to ensure seamless integration into the overall project.

3D printing in construction, an emerging technology, has allowed for the fabrication of building components or even entire structures. This reduces the need for traditional labor and material sourcing, enabling faster and more cost-efficient construction. Early experiments with 3D-printed homes and offices are paving the way for a more sustainable and rapid approach to construction.

Lean construction is a mantra for maximizing efficiency and reducing waste. Lean construction is a methodology that focuses on minimizing waste while maximizing value for clients. This involves optimizing processes, improving project delivery, and ensuring continuous improvement through feedback loops. The core principle of lean construction is to provide more value with fewer resources.

Adopt the "Last Planner System" to improve scheduling and reduce delays by involving all team members in planning and decision-making. Focus on continuous process improvement through feedback loops at every phase of the project. Reduce waste by reusing materials, optimizing supply chain logistics, and minimizing overproduction. The use of Artificial Intelligence (AI) in construction project management can support lean principles by predicting potential delays and resource shortages. AI tools can analyze historical project data and provide insights on areas where waste can be reduced, allowing project managers to make data-driven decisions.

Lean construction focuses on delivering maximum value by eliminating waste—whether that's time, materials, or effort—while continuously improving processes. Key tactics for industry leaders to implement lean principles include adopting the "Last Planner System" to involve all project participants in creating and updating schedules, ensuring that each phase aligns with the overall project timeline. Emphasizing continuous improvement through regular team meetings to discuss process efficiency and tackle bottlenecks helps. Developing strong supplier relationships to enhance material delivery systems and reduce delays is necessary.

The Stanford Hospital Expansion in California utilized lean construction principles to great success. By incorporating lean methodologies such as the "Last Planner System" and focusing on just-in-time delivery, the hospital's expansion was completed 12% faster than initially estimated, resulting in significant cost savings.

The project team also reported a 25% reduction in waste, thanks to the strategic approach to resource management and continuous process optimization.

I suggest the project leader hosts regular post-project reviews to capture lessons learned and continuously refine processes. Create cross-functional teams to focus on different aspects of waste reduction, such as material management, labor productivity, and logistics. Invest in training your teams on lean principles to ensure they understand the importance of minimizing waste and maximizing value.

Sustainable building practices are essential for green construction. Sustainability is no longer optional but a core requirement for modern construction projects. The focus has shifted from energy-efficient buildings to sustainable design principles including reducing the carbon footprint, reusing materials, and considering the building's full lifecycle.

Sustainability and green construction reduce environmental impact. Sustainability is no longer an afterthought in construction—it's a critical requirement. Industry leaders must prioritize green building techniques, such as energy-efficient designs, renewable energy integration, and sustainable materials. To drive sustainability in construction, adopt green building standards like LEED (Leadership in Energy and Environmental Design) and target net-zero energy consumption for new projects. Use sustainable materials like recycled steel, low-carbon concrete, and eco-friendly insulation to reduce the environmental footprint of each project. Implement energy-efficient systems during both the design and operational phases, such as solar panels, geothermal systems, and smart building technologies.

The Edge in Amsterdam is a shining example of sustainable construction. As one of the greenest office buildings in the world, it incorporates a combination of solar power, smart building technologies, and sustainable materials to reduce its environmental impact. The building's energy efficiency was optimized through its use of IoT devices that track energy usage in real-time and adjust lighting, heating, and cooling accordingly, resulting in a 70% reduction in energy consumption compared to traditional office buildings.

I firmly believe in making sustainability a key performance indicator (KPI) for all projects, tracking environmental impact from the early design phase through post-construction. Partner with green building certification organizations to ensure that projects meet or exceed the required sustainability standards. Explore renewable energy solutions, such as wind, solar, and geothermal, as viable energy options for your construction projects.

Design with energy efficiency in mind, including passive design elements such as natural ventilation, insulation, and renewable energy systems. Implement green construction techniques like using recycled materials, low-emission equipment, and reducing construction waste. Invest in energy-efficient systems like smart lighting, HVAC, and building management systems to reduce operating costs and environmental impact.

Smart cities and buildings are the future of urban development. Integrating IoT (Internet of Things) sensors into construction projects allows for real-time data collection and management, providing insights into energy consumption, building performance, and occupancy levels. These technologies enable adaptive building systems that optimize energy use and reduce operational costs.

Drones and aerial surveillance are being used extensively for monitoring and inspection across the globe. Drones are making a significant impact on construction projects by providing a cost-effective, accurate, and fast way to inspect sites, monitor progress, and gather real-time data. Drones can be used for surveying, capturing high-resolution images, and conducting site inspections.

Integrate drones into the site inspection process to quickly identify discrepancies or potential issues before they escalate. Use drones to create accurate topographic maps and 3D models of construction sites, reducing the time spent on manual measurements. Monitor construction progress remotely using drone footage, ensuring that the project stays on schedule and budget.

With the integration of AI and machine learning, drones are now capable of analyzing site data autonomously. This means that drones can not only capture images but also assess construction quality, identifying risks such as structural defects or inconsistencies in the construction process.

Robotics and automation enhance safety and precision. Robotics is beginning to play a major role in the construction industry, improving productivity and safety on job sites. From robotic bricklayers to autonomous bulldozers, robots can handle repetitive, hazardous tasks, reducing worker injury risks and increasing precision.

Invest in robotics for tasks that are repetitive or physically demanding, such as bricklaying, concrete pouring, and demolition. Train workers to collaborate with robots and use automated systems efficiently. Implement robotics in safety inspections to reduce human exposure to high-risk environments.

Exoskeletons, wearable robotic devices that assist workers by enhancing strength and reducing fatigue, are also emerging in the construction industry. These systems allow workers to handle heavy materials without straining their bodies, improving both productivity and safety.

Collaborative projects are delivery models in today's world of innovations and advancements. The traditional model of construction often places contractors, architects, and engineers in competing silos, which can lead to delays and cost overruns. Collaborative project delivery models, such as Integrated Project Delivery (IPD) and Design-Build, emphasize teamwork and shared risk/reward to improve outcomes.

Adopt collaborative project delivery models to create a more integrated, transparent environment. Involve all stakeholders early in the design and planning process, including subcontractors, to ensure alignment of goals. Use collaborative tools, such as shared cloud-based project management platforms, to ensure real-time communication and coordination across teams.

Blockchain technology is making its way into construction project management. By using blockchain, construction professionals can ensure that contracts are transparent and enforceable, payments are automated based on milestones, and all project data is secure and easily accessible to all stakeholders.

Knowledge about smart wearables and risk management technologies is essential for safety. Safety is a top priority in construction, and innovative technologies like smart wearables and AI-driven risk management systems are playing an important role in reducing on-site accidents and improving overall safety. Key steps to enhance safety on construction sites include equipping workers with wearable devices such as helmets with built-in sensors to track vital signs and detect dangerous situations, such as excessive heat or falls. Leverage AI and data analytics to predict potential safety hazards based on site conditions, worker behavior, and environmental factors. Implement real-time monitoring systems that send instant alerts to supervisors if hazardous conditions are detected.

Skanska, a global construction company, successfully integrated wearable safety technology in their projects. Their use of wearable sensors helped detect unsafe environmental conditions, such as extreme temperatures, and monitored the physical well-being of workers, reducing accidents by 30%. By using real-time data from wearables, they were able to respond to safety concerns before they escalated, enhancing overall worker safety.

It is sincerely advised to make safety technology a key investment area in your organization's strategic planning, and ensure that workers receive proper training on the use of wearable devices. Establish a safety-first culture by setting clear safety goals, incentivizing safe behavior, and integrating safety into every stage of the project lifecycle. Regularly audit your safety practices, using data from wearables and monitoring tools to identify areas for improvement.

Smart wearables available in the market are safety innovations. Safety is a primary concern on construction sites with workers exposed to numerous hazards. Smart wearables—such as helmets with integrated sensors, smart vests, and GPS tracking devices—are increasingly being used to monitor worker health and safety in real-time.

Implement wearable technologies to monitor worker vitals, location, and exposure to hazardous conditions (e.g., heat, noise, etc.). Use real-time data from wearables to alert managers to potential safety risks, allowing for quicker interventions. Incorporate safety training and education about the importance of wearables and personal protective equipment (PPE).

AI-powered safety systems that analyze wearable data and alert managers to potential risks or incidents before they happen are emerging. These systems can prevent accidents by predicting potential hazards based on real-time environmental and worker data.

The construction industry is evolving rapidly, driven by technological advancements, environmental challenges, and the need for greater efficiency. By embracing digital tools such as BIM, adopting modular construction methods, focusing on sustainability, and integrating cutting-edge innovations like drones, robotics, and AI, professionals can improve project delivery, reduce costs, and create safer, more sustainable buildings. The strategies outlined in this playbook can serve as a roadmap for construction professionals seeking to stay ahead of the curve and deliver successful projects in a competitive and ever-changing landscape.

The construction industry is facing a pivotal moment where technology, sustainability, and operational efficiency are becoming more critical than ever. Industry leaders who want to maintain a competitive edge and lead their organizations to success must embrace innovative strategies, tools, and best practices. This guide provides actionable insights, success stories, and tactics that can help construction professionals adapt to the changing landscape, deliver more efficient projects, and drive long-term success.

Data-driven decision-making can significantly enhance construction project outcomes, from improving resource management to predicting project risks. Implementing AI and machine learning into project management allows for predictive analytics, which can forecast project delays, cost overruns, and potential issues.

Turner Construction used predictive analytics to improve project outcomes on several high-profile projects. By analyzing historical project data and incorporating AI algorithms, Turner was able to predict project risks, allocate resources more effectively, and reduce delays by 20%. Their use of data-driven decision-making also helped reduce costs by optimizing the supply chain and workforce management.

I recommend investing in AI and machine learning tools that analyze large volumes of data to predict risks, costs, and timelines for construction projects. Use predictive analytics to enhance resource allocation, scheduling, and budgeting, ensuring that decisions are made based on actionable data. Develop a data-driven culture within your organization by encouraging teams to use real-time data for decision-making and problem-solving.

The future of construction lies in leveraging innovative technologies, sustainable practices, and a data-driven approach to project management. By implementing strategies such as BIM, lean construction, modular building, safety technology, and AI-powered analytics, industry leaders can enhance operational efficiency, reduce costs, and deliver better-quality projects. Success stories such as Hudson Yards, the Stanford Hospital Expansion, and The Edge exemplify how adopting these innovations can lead to groundbreaking results. Leaders who embrace these strategies will be better positioned to tackle the evolving demands of the construction industry while creating a safer, more sustainable, and efficient future.

Autonomous Paving, Robotics, and 3D Paving Resources go hand in hand. There are many real-world examples. China's unmanned road resurfacing is a prominent example. China has made significant strides in fully autonomous road resurfacing. Notably, they've completed projects using fleets of autonomous pavers, rollers, and drones. These projects demonstrate the potential for large-scale, fully automated road construction.

It is worth mentioning the Beijing-Hong Kong-Macao Expressway Resurfacing project. China completed the world's first fully unmanned paving construction along a 157.79 km stretch of the Beijing-Hong Kong-Macao Expressway using autonomous machinery and drones. The project included autonomous equipment, precision technology, innovative paving, and safety features.

A fleet of autonomous road equipment supplied by Sany, including a 20-meter-wide unmanned paver, six 13-tonne double-drum rollers, and three 30-tonne rubber-wheel rollers operating in a synchronized formation. The Beidou satellite positioning system provided centimeter-level precision, while advanced algorithms and a low-latency communication network enabled real-time path planning and operational control. The SAP200C-10 paver achieved a world-first by laying a 19.25-meter-wide stretch of asphalt in a single pass, reducing joints and improving road smoothness. Multiple safety layers including electronic fences, emergency stop functions, and real-time obstacle detection systems ensured safe operations and reduced human presence in hazardous environments. It is being followed by many countries now.

The development of self-driving pavers that can precisely lay asphalt with minimal human intervention is a major focus of autonomous paving. Autonomous rollers are also being implemented to ensure consistent compaction.

Similarly, InfraROB Project in Europe is a breakthrough in this field. The InfraROB project, involving multiple European countries, is focused on developing and testing robotic solutions for road construction and maintenance. This project has featured successful trials of fully autonomous asphalt paving, robotic placement of safety barriers, and autonomous road marking.

Strabag's Autonomous Paving Trials is another significant example. Strabag has been conducting trials of autonomous road paving technology, demonstrating the practical application of these advancements. Their work on the A9 motorway in Austria is a great example of this and is worth knowing about.

The A9 Motorway Renovation project by STRABAG successfully tested fully autonomous asphalt paving on an 180-meter-long section of the A9 motorway between Graz and the Slovenian border as part of the EU-wide InfraROB research project. Autonomous paving, sensor integration, and focus on safety are some of the major highlights of the innovate project.

A fully autonomous paver laid asphalt completely automatically, with implementation and logistics coordinated by a STRABAG team. A sensor-equipped fiber optic cable was integrated into the asphalt surface during paving to provide information about load on the asphalt surface during operation. Mobile safety cone robots with object recognition sensors managed traffic closure, while small autonomous robots marked the road after paving. The project aimed to improve occupational safety and reduce physical strain on workers by keeping paving personnel at a greater distance from traffic flow.

These projects demonstrate how robotics, automation, and advanced paving technologies are transforming the asphalt paving industry by improving efficiency, quality, safety, and

Thus, 3D Printing is majorly being used for road repair. Robots equipped with 3D printing capabilities are being developed to automate pothole and crack repair. They are being used for tasks like placing safety barriers, road marking, and site surveying for better site management. A primary driver for automation is to improve worker safety by reducing their exposure to hazardous conditions for increased safety and efficiency, especially of human resources. Automation also aims to increase efficiency and consistency in paving operations.

Robotic 3D printing for road repair projects focus on the integration of sensors into the road surface during the paving process, to allow for better monitoring of the road's condition.

To sum up, I can say that the use of robotics and automation in asphalt paving is rapidly evolving. These technologies have the potential to significantly improve safety, efficiency, and quality in road construction. Ongoing research and development are paving the way for wider adoption of these technologies in the future.

Talking about risks, an innovation I led was the design aimed at mitigating distracted driving risks in active construction zones. **My Company's Director of Safety, Todd Atkins, approached me with this problem.** Recognizing the hazard posed by drivers unknowingly entering dangerous work areas — especially when using mobile devices — I conceptualized a solution that leverages GPS-based geofencing to trigger audio-visual alerts on smartphones as they approach designated zones.

The system includes features for both drivers and supervisors, allowing for real-time speed tracking, distance measurement from work zones, and even unsafe zone reporting. Its goal is to actively prevent accidents through situational awareness, while being simple enough to deploy without requiring new vehicle hardware.

This project reflects my ongoing mission to bridge digital technology and field-level safety, especially in environments as complex and fast-moving as road construction.

To address the challenges of limited visibility and human error near paving machinery, I also developed a prototype system using computer vision and edge computing to detect and respond to safety hazards in real-time.

Using object detection models (such as YOLO) and compact AI hardware (like Jetson Nano), the system can identify nearby workers, safety cones, or vehicles — and then autonomously trigger safety responses such as alerts or equipment slowdowns. It's designed to work without cloud connectivity, operating independently in the rugged, data-limited environments common on job sites.

This tool represents an important shift toward autonomous safety response in the construction industry, and reflects my broader vision of embedding intelligence into the equipment that shapes our infrastructure.

Another significant innovation that is the future of road construction is the use of nano-enhanced asphalt. As global traffic grows and infrastructure demands increase, the need for durable and sustainable road surfaces has never been more pressing. Traditional asphalt, while functional, often suffers from issues such as cracking, deformation, and weathering over time, leading to costly repairs and a shorter lifespan. To address these challenges, researchers and engineers are turning their attention to innovative materials, particularly nano-enhanced asphalt. This advanced approach involves the incorporation of nanoparticles, such as carbon nanotubes and silica, into asphalt mixes to significantly improve their properties.

At the core of nano-enhanced asphalt is the concept of nanotechnology, which deals with materials at the atomic and molecular scale. By integrating nanoparticles into asphalt, the overall structure and performance of the material can be transformed. Carbon nanotubes, for example, are known for their exceptional strength and flexibility. When added to asphalt, they create a composite material that exhibits enhanced mechanical properties. This not only increases the tensile strength of the pavement, allowing it to withstand heavier loads and stress, but also improves its flexibility, reducing the likelihood of cracking under temperature fluctuations.

Another benefit of incorporating silica nanoparticles into asphalt is their ability to enhance resistance to weathering. Traditional asphalt is susceptible to degradation from environmental factors such as UV radiation, moisture, and temperature changes. Silica, with its high surface area and reactive properties, can help to create a more robust matrix that shields the asphalt binder from these harmful effects. This enhancement means roads can maintain structural integrity for a longer period, ultimately leading to a decrease in necessary maintenance and repair work.

The benefits of nano-enhanced asphalt extend beyond just strength and durability. The environmental impact of road construction and maintenance is a growing concern as well. By creating pavements that last longer and require less frequent repairs, the overall carbon footprint associated with road work can be significantly reduced. In addition, the use of nanoparticles can lead to more efficient use of materials, minimizing waste during the production and application of asphalt. This is particularly important in the context of sustainability as societies strive to meet growing infrastructure needs while also addressing environmental challenges.

Furthermore, the incorporation of nanotechnology in asphalt can lead to smart pavement solutions. For instance, researchers are exploring the possibility of embedding sensors within nano-enhanced asphalt to monitor road conditions in real-time. These sensors could provide data on temperature changes, stress levels, and wear and tear. Such information could help authorities manage road maintenance more proactively, thereby enhancing safety and reducing costs.

Despite the promising advantages of nano-enhanced asphalt, there are still challenges to overcome. The production costs associated with developing and implementing these advanced materials can be significant, which might hinder widespread adoption. Moreover, the long-term effects and potential environmental implications of using nanomaterials in construction need thorough investigation. Regulatory frameworks will need to adapt to address these concerns, ensuring safety and environmental protection.

In conclusion, the integration of nanoparticles into asphalt represents a transformative approach to road construction. With the potential to enhance strength, flexibility, and resistance to weathering, nano-enhanced asphalt can lead to longer-lasting pavements with reduced maintenance needs. As research continues and technology advances, this innovative material has the power to reshape our infrastructure for the better, contributing to more sustainable and durable road systems that can meet the demands of modern society. Embracing these advancements could ultimately pave the way for smarter, stronger, and more resilient road networks globally.

CHAPTER 4

THE SMART SITE: TRANSFORMING CONSTRUCTION WITH TECHNOLOGY

In my early research efforts to improve infrastructure management, I explored the use of machine learning algorithms to forecast pavement degradation and optimize maintenance schedules. This work laid the foundation for predictive tools that can estimate the future condition of asphalt surfaces using historical performance data.

By integrating supervised models like **XGBoost** and time-series models such as **LSTM**, I demonstrated how road agencies and contractors could shift from reactive repairs to proactive asset management. The models were designed to process PCI (Pavement Condition Index) trends over time, factoring in environmental and load-related stresses.

This research not only reduced guesswork in maintenance planning, but also introduced a scalable framework for applying AI in road infrastructure — a critical step toward data-driven public works.

The construction industry has long been characterized by its traditional practices and methods, but recent advancements in digital tools and smart technology are reshaping the landscape. The integration of these innovations is not only revolutionizing the way structures are designed and built, but also enhancing efficiency, safety, and sustainability across the sector.

One of the most significant trends in contemporary construction is the advent of Internet of Things (IoT)-enabled buildings. These structures are equipped with sensors and devices that collect real-time data, allowing for better monitoring and management of various operations. For instance, smart HVAC systems can automatically adjust parameters based on occupancy levels and outdoor weather conditions, leading to optimized energy consumption. This lowers operational costs for building owners and contributes to environmental sustainability by reducing buildings' carbon footprints.

Moreover, IoT technology facilitates predictive maintenance. By continuously monitoring equipment and infrastructure, potential issues can be identified before they escalate into costly repairs or replacements. This proactive approach minimizes downtime and extends the lifespan of building systems, providing a clear advantage over traditional maintenance practices that often rely on scheduled check-ups.

In addition to IoT, artificial intelligence (AI) is making waves in project management within the construction industry. AI-driven solutions analyze vast amounts of data to identify patterns and insights that can help optimize workflow. For instance, AI can assist in resource allocation by predicting labor needs based on historical project data. By leveraging machine learning algorithms, construction managers can make more informed decisions, ultimately leading to improved project timelines and budget adherence.

Another area where AI is proving invaluable is risk management. Construction projects are inherently risky, with multiple variables that can impact the outcome, from weather conditions to supply chain disruptions. AI tools can analyze this data in real-time, providing insights that help project managers mitigate potential risks, enabling them to adapt strategies swiftly and effectively.

Moreover, the integration of Building Information Modeling (BIM) with AI presents exciting possibilities. BIM allows for the creation of digital representations of physical and functional characteristics of places. When paired with AI, these models can simulate different scenarios, helping teams visualize the impact of certain decisions before they are made. This capability enhances collaboration among stakeholders and ensures that potential challenges are addressed during the planning stages, resulting in fewer surprises during construction.

As the construction industry continues to evolve, the emphasis on sustainability and resource efficiency is becoming paramount. Smart technologies play a crucial role in achieving these goals. For example, by implementing energy management systems that utilize real-time data, construction firms can create greener buildings responsive to their environments. This trend aligns with a broader societal push towards sustainable development, as consumers and investors increasingly prefer eco-friendly practices.

In conclusion, the integration of digital tools and smart technology in construction is transforming the industry in profound ways. From

IoT-enabled buildings that optimize energy usage to AI-driven project management methods that enhance decision-making, these advancements are paving the way for a more efficient, safe, and sustainable future. As these technologies continue to evolve, we can expect even more innovative solutions that will redefine the way we build and manage our environments, marking a significant shift towards a smarter, more connected construction industry.

The condition of roads across the United States is steadily deteriorating, as highlighted by the "D" grade issued in the 2021 American Society of Civil Engineers (ASCE) Infrastructure Report Card. Major factors contributing to this decline include aging infrastructure, insufficient funding, and growing traffic demands. These issues not only affect road quality but also drive the need for innovation in road maintenance. One of the biggest challenges for road authority today is the difficulty of accurately identifying which roads require repairs and their exact locations. Existing methods for assessing road conditions, even when adhering to standardized protocols, fall short. However, advancements in technology, particularly artificial intelligence (AI) and machine learning (ML), offer promising opportunities to address these challenges and enhance road assessment techniques as mentioned earlier.

Managing 4 million roads in the United States poses a significant challenge, making it infeasible for humans to inspect each one frequently to identify specific issues. With the help of AI, Michelin Better Roads provides communities with a comprehensive overview of their entire road network's condition, allowing for proactive pavement maintenance planning that helps prevent the formation of potholes and other serious damage. Since its development at Carnegie Mellon University in 2016, RoadBotics—now Michelin Better Roads—has utilized proprietary machine learning algorithms to evaluate more than 49,000 miles of road, establishing itself as a leader in automated road assessments.

The integration of artificial intelligence (AI) and machine learning in asphalt paving is transforming the construction industry by enhancing efficiency, quality, and safety. Traditionally, asphalt paving has been a labor-intensive process, often reliant on manual oversight and experience. However, advancements in technology are paving the way for data-driven methodologies that optimize every step of the paving process.

One primary application of AI in asphalt paving involves the analysis of data collected from various sources, such as sensors, drones, and real-time monitoring systems. These technologies can collect data on material properties, environmental conditions, and equipment performance. Machine learning algorithms can sift through this vast amount of information to identify patterns, allowing project managers to make informed decisions about material selection, temperature control, and application techniques.

For example, temperature control is critical during the asphalt laying process. If the material is too hot when applied, it can lead to defects, while applying it too cold can result in poor bonding with the underlying layers. AI systems can predict the optimal temperature ranges based on environmental factors and adjust the equipment settings accordingly. This proactive approach enhances the durability of the paved surface and reduces the likelihood of costly repairs down the line.

Additionally, AI and machine learning contribute to improving safety on construction sites. Automated systems equipped with AI are capable of monitoring personnel movements and equipment operations, thereby enhancing situational awareness. These systems can detect potentially hazardous conditions, such as equipment crowding or worker proximity to moving vehicles, alerting operators to take corrective actions before accidents occur.

Moreover, the predictive capabilities of machine learning allow for better maintenance scheduling. By analyzing historical performance data and predicting future conditions, paving authority can proactively address maintenance needs, extending the lifespan of asphalt surfaces and minimizing disruptions to traffic.

Finally, the integration of AI and machine learning facilitates more sustainable practices in asphalt paving. By optimizing material usage and minimizing waste, the environmental impact of paving projects can be significantly reduced. For instance, AI algorithms can help recycle old asphalt more efficiently, promoting sustainability through circular construction practices.

In summary, the incorporation of AI and machine learning in asphalt paving is not just a trend but a significant evolution in the industry. By improving efficiency, safety, and sustainability, these technologies represent a smart investment for paving companies looking to enhance their operations and deliver high-quality results. As the technology continues to advance, we can expect even greater improvements in how asphalt paving projects are planned and executed.

The integration of machine learning into the maintenance and optimization of asphalt pavements represents a progressive advancement in civil engineering and infrastructure management. Traditionally, the maintenance of roadways has relied on periodic inspections, experience, and heuristic approaches to predict when repairs are needed. However, the growing complexity of transportation systems, coupled with the increasing demand for efficiency and sustainability, necessitates a more data-driven approach. Utilizing machine learning algorithms is one of the most promising strategies to meet these challenges.

At the core of this innovative strategy is the deployment of sensors embedded in asphalt pavements. These sensors can continuously monitor various conditions, collecting data on factors such as temperature, moisture levels, traffic loads, and even structural integrity. This real-time data provides a comprehensive view of pavement conditions, allowing for a more nuanced understanding of their performance over time. By harnessing this wealth of information, machine learning algorithms can analyze and identify patterns that may not be readily apparent through manual inspections.

One of the most significant advantages of employing machine learning in pavement maintenance is the predictive capability it offers. Predictive maintenance is a proactive approach to managing infrastructure assets, as opposed to reactive maintenance which often occurs after a problem has already manifested. By processing historical data along with real-time sensor readings, machine learning models can forecast when a pavement will require maintenance or repairs. This capability is particularly valuable because it allows for timely interventions that can prevent more extensive and costly damage.

For instance, if a machine learning model identifies that particular conditions, such as sustained heavy rainfall followed by increased traffic loads, correlate with expedited pavement deterioration, maintenance can be scheduled during anticipated stress periods. This kind of foresight not only extends the life of the pavement but also reduces the overall maintenance costs, as repairs can be carried out at optimal times, when they are least disruptive to traffic and the public.

Moreover, the optimization of repair schedules through machine learning leads to enhanced efficiency in resource allocation. Instead of performing maintenance based solely on a schedule or fixed timeframe, maintenance teams can prioritize repairs based on actual conditions and needs identified through data analysis. This prioritization ensures that resources are used where they are most needed, which is particularly important in scenarios where budgets and labor are constrained.

Another aspect to consider is the potential for machine learning algorithms to analyze and improve the overall lifecycle of materials used in pavement construction and maintenance. Data from sensors can provide insights into how different pavement materials perform under various conditions, leading to recommendations for more durable materials or innovative construction techniques that optimize performance and longevity. This continuous improvement approach can support the development of smarter materials that adapt to changing environments and traffic conditions.

The collaborative capabilities of machine learning also promote enhanced stakeholder engagement. Municipalities, transportation agencies, and private contractors can benefit from dashboards or visualization tools that present real-time data in an accessible format. Decision-makers can visualize trends, costs, and predictive insights at a glance, facilitating more informed decisions about budget allocation, project prioritization, and long-term strategic planning. This transparency can help foster collaboration among various stakeholders, aligning efforts toward the common goal of maintaining efficient and durable roadways.

However, successfully implementing machine learning for pavement management does come with its challenges. One of the primary hurdles is data quality and standardization. To develop effective predictive models, high-quality data must be collected consistently across different locations and environmental conditions. If data is inconsistent or incomplete, it can compromise the accuracy of the predictions. Therefore, establishing standardized protocols for sensor installation, data collection, and processing is crucial for the success of this initiative.

Additionally, there is the need for ongoing training and validation of machine learning models. As new data becomes available, models must be updated and validated to ensure they remain accurate and reliable. This requires a commitment to ongoing collaboration between data scientists, engineers, and field personnel to continually refine the algorithms based on real-world performance metrics.

Another consideration is the integration of this technology into existing systems. Transitioning to a data-driven maintenance strategy involves ensuring that current infrastructure and workforce are equipped to handle these new tools. This may require retraining staff or even redefining roles within maintenance teams. However, the long-term benefits—such as reduced costs and a more dependable roadway network—can justify these initial investments.

In conclusion, the application of machine learning algorithms to analyze data from sensors embedded in asphalt pavements heralds a new era in pavement management. By predicting maintenance needs and optimizing repair schedules, this approach not only reduces maintenance costs but also enhances the longevity of pavements. The combination of real-time data collection, predictive analytics, and efficient resource allocation creates a responsive and proactive maintenance strategy, ultimately leading to safer, more durable roadways. As this technology continues to evolve, it holds the promise of transforming infrastructure management, making it more efficient, effective, and sustainable for future generations.

The research titled "Prediction of Moisture Susceptibility of Asphalt Mixtures Containing RAP Materials Using Machine Learning Algorithms," published in the International Journal of Pavement Engineering, examines how machine learning (ML) algorithms can forecast the moisture susceptibility of asphalt mixtures that include reclaimed asphalt pavement (RAP) materials. The study assessed four ML techniques—decision tree (M5P), random forest (RF), gradient boosting (GB), and extreme gradient boosting (XGBoost)—to evaluate their effectiveness in enhancing pavement durability. The results indicated that ML models can effectively predict moisture damage in asphalt mixtures, enabling engineers to optimize material choices and evaluate the risk of pavement failure over time. Notably, the XGBoost model demonstrated the best predictive performance, achieving a correlation coefficient of 0.96 for training data and 0.75 for testing data. These findings underscore the potential for ML-driven predictive models to strengthen asphalt pavement durability. Furthermore, comparisons between actual and predicted tensile strength ratio (TSR) values, along with residual analysis, validated the reliability and robustness of these models.

The implications of this study are particularly pertinent given the challenges facing road maintenance in the United States. In 2021, state and local governments allocated over $206 billion for road upkeep, while the Department of Transportation reported a $1 trillion backlog of bridge and road repairs in 2023. By enhancing asphalt mixtures through ML approaches, it's possible to lower these expenses and prolong the lifespan of roadways.

Traditionally, determining the optimal mix of Recycled Asphalt Pavement (RAP) and other materials for various weather conditions involves lengthy and expensive laboratory tests. However, AI-driven models provide a quicker and more cost-effective solution, making infrastructure development more economical and environmentally sustainable. This research highlights the potential of machine learning models to improve the accuracy and efficiency of predictions regarding moisture susceptibility, which ultimately supports the creation of more durable asphalt pavements. The findings can help engineers, transportation departments, and private sector stakeholders implement more sustainable and cost-effective design strategies.

As the demand for safer and more efficient transportation infrastructure grows, the integration of advanced technology into our road systems has taken on increased significance. One promising innovation in this field is the development of advanced sensor networks embedded directly in asphalt. These networks are designed to continuously monitor critical road conditions—including temperature, traffic load, and material stress. By leveraging real-time data gathered from these sensors, transportation agencies and municipalities can implement predictive maintenance strategies, thereby minimizing costly repairs and extending the lifespan of pavement.

Roadways are subject to a variety of environmental and mechanical stresses. Factors such as temperature fluctuations, the weight of vehicles, and the wear and tear from constant usage can lead to pavement deterioration over time. Traditionally, road maintenance has relied on periodic inspections and rough assessments, which often do not accurately capture the current state of the pavement and its underlying structure. This reactive approach can lead to more severe damage by the time repairs are initiated, resulting in higher repair costs and safety risks.

Incorporating sensor technologies into the pavement itself allows for a paradigm shift in how road conditions are monitored. Instead of waiting for visible signs of distress or scheduling routine inspections, real-time data collection enables proactive management of road infrastructure. This continuous monitoring provides decision-makers with critical insights into when and where maintenance is needed, ultimately improving road safety and reducing overall costs.

It is important to know the types of sensors and their functions for optimum use. The success of embedded sensor networks relies on the deployment of several sensor types, each designed to measure specific road conditions:

1. Temperature Sensors: These sensors provide real-time data on the temperature of the asphalt. Temperature fluctuations can significantly affect the material properties of asphalt, such as its viscosity and tendency to crack. By monitoring temperature

conditions, maintenance teams can better predict when softening or cracking may occur and plan interventions accordingly.

2. Traffic Load Sensors: Traffic load sensors assess the weight and frequency of vehicles traveling over the road. This data is crucial for understanding the stress levels that pavement endures and anticipating when materials may begin to yield. Accumulating data on traffic patterns allows for better planning of maintenance schedules to align with peak load periods.

3. Material Stress Sensors: These sensors gauge the internal stresses within the asphalt due to factors such as temperature changes and traffic loads. Understanding material stress helps in identifying potential structural failures before they manifest as physical damage on the surface.

4. Moisture Sensors: Embedded sensors can measure moisture levels within the asphalt. Excess moisture can lead to significant degradation of road materials and can accelerate cracking and pothole formation. Monitoring moisture levels allows maintenance teams to address drainage issues promptly and improve overall pavement durability.

Once the sensors are embedded in the asphalt, they continuously gather data and transmit it to a central analytics platform. Data can be processed using advanced algorithms and machine learning techniques to detect patterns and predict when maintenance should take place. For instance, algorithms can analyze traffic patterns alongside material stress and temperature data to predict potential failures or the optimal timing for resurfacing.

The integration of this data with Geographic Information Systems (GIS) can also provide visualizations that help in strategic decision-making. Interactive maps can highlight areas of concern, allowing transportation officials to prioritize maintenance tasks based on real-time conditions rather than relying solely on historical data.

Predictive maintenance represents a transformative approach to road management. By utilizing the data collected from sensor networks, transportation agencies can shift their focus from reactive repairs to proactive maintenance. This can lead to several beneficial outcomes like cost savings, reduced disruption, enhanced safety and sustainable practices.

By identifying issues before they escalate into major problems, agencies can save on repair costs and extend the life of the pavement. Proactive interventions, such as targeted resurfacing or crack sealing, can be significantly more cost-effective than extensive repairs.

Targeted maintenance efforts can minimize traffic disruptions. Maintenance teams can schedule work during off-peak hours and focus their efforts on specific areas that require attention, rather than conducting widespread repairs that affect traffic throughout the entire region.

Continuous monitoring leads to timely repairs, which can reduce the risk of accidents associated with road failures. Keeping roads in optimal condition contributes to overall transportation safety for all road users.

Prolonging the life of existing pavement through effective maintenance practices contributes to more sustainable infrastructure management. It reduces the need for new materials and minimizes the environmental impact associated with road construction.

While the development of embedded sensor networks offers numerous advantages, some challenges and considerations must be addressed. The initial installation of sensor networks can be costly, requiring significant investment in both technology and infrastructure. However, the long-term savings potential can justify this cost. The sheer volume of data generated by sensor networks necessitates robust data management strategies. Developing efficient systems for data storage, analysis, and visualization is critical to maximizing the benefits of the technology. For many regions, integrating new sensor networks with existing infrastructure management systems can pose challenges. Ensuring compatibility and interoperability is crucial for seamless operations. The sensors themselves require maintenance and periodic calibration to ensure accuracy. Agencies must develop strategies to manage and maintain the sensor networks throughout their lifespan.

The integration of advanced sensor networks into asphalt presents a revolutionary opportunity for improving road maintenance practices. By enabling continuous monitoring of critical road conditions, these networks provide invaluable data that can inform predictive maintenance strategies. In doing so, they help to minimize repair costs, enhance safety, and ultimately extend the lifespan of our transportation infrastructure.

As cities and municipalities continue to invest in smart infrastructure, the use of embedded sensor networks will likely become a standard practice. With the potential to improve efficiency and sustainability in road management, this technology represents a significant step forward in the quest for safer, smarter, and more resilient transportation networks.

Smart parking systems are the demand of every city. The integration of sensors, cameras, and software in parking lots represents a significant advancement in urban mobility and infrastructure. As cities continue to grow and the number of vehicles increases, parking has become a critical concern, leading to congested streets and frustrated drivers. Traditional parking methods often involve circling to find an available spot, which not only wastes time but also contributes to increased emissions and reduced air quality. Thus, the implementation of technology in parking lots can lead to considerable improvements in space efficiency, reduce congestion, and enhance overall customer satisfaction.

With urban areas experiencing exponential growth, the challenge of parking has reached new heights. Increasingly crowded city streets, coupled with limited parking infrastructure, have created a perfect storm for parking-related issues. According to studies, drivers can spend an average of 20 minutes looking for parking in congested areas, leading to wasted time and increased frustration. This situation has prompted a need for innovative solutions that streamline parking processes and improve the overall experience for drivers.

The integration of sensors, cameras, and sophisticated software systems into parking lots can facilitate real-time detection of available parking spaces. This technology works on several fronts, employing various tools like sensors, cameras and software solutions to streamline the parking process.

These devices can be installed in each parking space to determine whether it is occupied or vacant. Ultrasonic sensors are often used, sending sound waves to detect vehicles within their vicinity. When a vehicle parks in the designated spot, the sensors detect the change and send this information to the central software system. Video surveillance cameras can also contribute to parking management by continuously monitoring the parking area. Advanced image processing technology enables these cameras to recognize available spaces and track vehicle movements. They can also provide valuable data regarding peak usage times, types of vehicles parked, and patterns that can be analyzed for future improvements. Central to the integration of these technologies is robust software that processes data from sensors and cameras. This software can be configured to notify drivers through mobile applications, digital signage, or in-car navigation systems about available spaces. It can also offer features such as reservation systems and payment processing, making the entire parking experience more seamless.

The integration of technology into parking lots offers numerous benefits like space efficiency, reduced congestion, enhanced customer satisfaction, environmental impact, and data collection and insights.

With real-time data, parking lots can maximize their capacity efficiently. By reducing the time vehicles spend searching for spots, the overall use of parking spaces is optimized. Additionally, better management can lead to the redesign of parking lots to better accommodate the flow of traffic and increase the number of vehicles that can be parked.

By guiding drivers directly to available parking spaces, the technology can significantly reduce the volume of traffic in and around parking structures. This reduction can lead to lower emissions and a decrease in road congestion, as vehicles spend less time idling or maneuvering in search of a parking spot.

The modern driver values convenience and efficiency. By providing real-time updates on parking availability, drivers can experience reduced stress and frustration associated with finding parking. Furthermore, features like advanced reservations can ensure that users have a guaranteed spot upon arrival, further enhancing their satisfaction.

Smart parking systems contribute to sustainability initiatives by reducing the overall carbon footprint of parking operations. Less driving time means reduced vehicle emissions and better space management means more efficient land use, which is critical in urban planning.

The systems provide a wealth of data that can be analyzed for various insights. This information can help city planners and businesses understand parking trends, peak usage times, and user preferences. By harnessing this data, cities can make informed decisions about future parking strategies and even urban layout.

While the advantages are compelling, several challenges accompany the introduction of these technologies. Upfront costs for installing sensors, cameras, and software systems can be significant. Budget constraints may deter some municipalities or businesses from pursuing these solutions. To mitigate costs, collaboration between public and private sectors may be necessary.

Ensuring that technology functions reliably and securely is paramount. System failures, network issues, or cyber vulnerabilities could disrupt operations. Rigorous testing, regular maintenance, and robust cybersecurity measures are crucial for sustaining a reliable system. Drivers accustomed to traditional parking methods may need time to adapt to new technologies. Public education campaigns can help ease the transition and demonstrate the benefits of smart parking solutions. The collection of data raises questions about privacy and surveillance. Users should be assured that their personal information is protected and that data will be used responsibly, with transparency about how it's collected and shared.

As the world becomes increasingly digitized, the future of parking will likely rely on continuous advancements in technology. Innovations such as artificial intelligence (AI) could further enhance the systems by predicting parking demand, optimizing availability, and integrating with broader transportation networks (including public transit).

Moreover, collaboration with vehicle manufacturers can lead to even more integrated solutions, where vehicles communicate with parking infrastructure to find and reserve spaces automatically. The potential for innovations like electric vehicle (EV) charging stations integrated with parking solutions could also cater to the growing demand for sustainable transportation options.

In conclusion, the integration of sensors, cameras, and software into parking lots represents a transformative shift in how we approach urban mobility. This technology not only addresses existing challenges of parking inefficiencies and congestion but also leads to a more sustainable and enjoyable experience for drivers. With continued advances and thoughtful implementation, smart parking solutions have the potential to redefine urban parking and contribute significantly to the development of smart cities. Ultimately, the goal should be to create systems that enhance efficiency while prioritizing the needs and satisfaction of users, paving the way for smarter, more sustainable urban environments.

CHAPTER 5

FOUNDATIONS OF CHANGE: REDEFINING THE CONSTRUCTION INDUSTRY

Predictive maintenance is a proactive approach that leverages real-time data, advanced analytics, and Internet of Things (IoT) technologies to anticipate equipment failures before they occur. This strategy relies on the use of sensors and condition monitoring techniques, such as vibration and oil analysis, to continuously evaluate the state of equipment. Employing machine learning algorithms and predictive analytics will generate actionable insights that enable maintenance teams to address potential issues before any equipment failure or unplanned downtime.

I will talk about another research work done by me on 'Smart Pavement Technology Using IoT-Enabled Sensor Networks'. In a parallel innovation stream, I investigated how embedded IoT sensor networks could transform conventional pavement into intelligent infrastructure. The concept involved integrating low-power sensors within roadways to monitor real-time variables like strain, temperature, water infiltration, and traffic loads.

These smart sensors, when paired with edge computing, enabled the system to detect early signs of deterioration and trigger predictive maintenance alerts. I also studied energy-harvesting mechanisms (such as vibration-based power) to allow for self-powered sensor nodes, enhancing sustainability and reducing maintenance needs.

The broader vision was to establish a foundation for autonomous infrastructure health monitoring, where the road itself participates in its own maintenance — advancing the idea of resilient, self-aware cities.

Before we delve further, it is essential to comprehend the basic difference between preventive maintenance and predictive maintenance. Preventive maintenance is scheduled based i.e. adhering to time intervals or usage, similar to routine automotive maintenance. On the other hand, predictive maintenance is condition-based and driven by real-time data monitoring rather than adhering to a predetermined schedule. This fundamental difference enhances the efficiency of maintenance practices and significantly reduces unnecessary maintenance activities, thus lowering overall maintenance costs.

The predictive maintenance market is undergoing substantial growth, propelled by the advancing capabilities of artificial intelligence and machine learning. These technologies are becoming increasingly sophisticated, resulting in more accurate predictions and streamlined maintenance processes. According to Next Move Strategy Consulting, the global predictive maintenance market, valued at 4.5 billion U.S. dollars in 2020, is anticipated to reach 64.3 billion U.S. dollars by 2030. This remarkable growth underscores the increasing adoption of predictive maintenance practices across diverse industries.

In this chapter, I shall explore modern construction's foundations: Technology, Leadership, and Sustainability. The construction industry is undergoing a transformative journey driven by the convergence of technology, effective leadership, and sustainability. These elements are becoming the essential pillars, much like the foundations of a building that support the modernization of construction practices. In this exploration, I will delve into how each of these components interconnects and redefines the way we approach construction in the 21st century.

Technology plays an imperative role in preparing a solid foundation. In modern construction, technology serves as a cornerstone that underpins efficiency, safety, and innovation. Technology is reshaping every stage of the construction lifecycle, from Building Information Modeling (BIM) to automated construction processes. BIM is one of the most significant advancements in construction technology. It allows the creation of detailed 3D models that integrate all aspects of a building's lifecycle—from design and construction to operation and maintenance. This enhances collaboration among stakeholders and minimizes errors, reducing waste. By visualizing the entire project in advance, teams can anticipate challenges and mitigate risks before they arise, ensuring a smoother construction process.

Automation and robotics in construction have the potential to revolutionize the industry by enhancing productivity and safety. Robots can take on dangerous tasks, such as bricklaying or demolition, reducing the risk of accidents on job sites. Additionally, automated machinery can work continuously, increasing efficiency while minimizing labor costs. As these technologies become more advanced and affordable, we expect to see a growing trend towards robotic-assisted construction.

Incorporating green technologies into construction practices is another technological advancement that contributes to sustainability. Innovations such as energy-efficient materials, solar panels, and smart building systems are becoming more prevalent. These technologies reduce the environmental impact of construction and improve the long-term value of properties by lowering operational costs and attracting environmentally-conscious consumers.

The rise of big data and analytics in construction is transforming project management and decision-making processes. With access to data from various sources, such as equipment usage, labor productivity, and material costs, construction managers can make informed decisions that optimize resources and reduce waste. Predictive analytics can also anticipate potential delays and budget overruns, allowing for the implementation of proactive measures.

While technology provides the tools for innovation, effective leadership guides the adoption and implementation of these changes. Leaders in the construction industry play a critical role in fostering a culture of innovation, collaboration, and sustainability.

Leaders must have a clear vision for the future of construction, emphasizing the importance of adopting new technologies and sustainable practices. By articulating this vision, they inspire teams to embrace change and explore alternative methods. A strong leader envisions a modernized industry and actively champions the strategies needed to achieve it.

The construction industry is inherently collaborative, requiring input from various stakeholders, including architects, engineers, contractors, and clients. Effective leaders promote a culture of collaboration where diverse perspectives are valued. By utilizing collaboration tools and encouraging open communication, leaders can enhance teamwork and ensure alignment across all parties involved in a project.

As the industry evolves, so must the skill sets of the workforce. Leaders should prioritize training and development initiatives that equip employees with the knowledge needed to leverage new technologies. This investment not only enhances employee satisfaction and retention but also ensures that teams are well-prepared to navigate the complexities of modern construction.

A culture of innovation is essential for modernizing construction practices. Leaders can encourage creativity by allowing teams to experiment with new ideas and technologies. Leaders can foster a mindset that embraces continuous improvement by creating an environment where employees feel safe to take risks and learn from failures.

Most importantly, we need to build for the future, and this underscores the Importance of sustainability. As concerns about climate change and environmental degradation continue to rise, sustainability has become a critical focus for the construction industry. By adopting sustainable practices, the industry can reduce its environmental footprint while creating structures that are energy-efficient and environmentally friendly.

Using sustainable materials, such as recycled steel, bamboo, and reclaimed wood, can significantly reduce the environmental impact of construction. Additionally, reducing waste on job sites, recycling materials, and using eco-friendly finishes contribute to a more sustainable approach. The integration of these practices not only benefits the planet but also appeals to consumers who prioritize sustainability.

Incorporating energy-efficient designs and technologies is paramount in modern construction. Features such as passive solar design, energy-efficient HVAC systems, and smart building technologies can significantly reduce energy consumption in buildings. By focusing on energy efficiency, construction firms can help reduce greenhouse gas emissions and meet regulatory requirements related to sustainability.

Sustainable construction goes beyond just materials and technologies; it also involves engaging with the community. Construction projects should consider the social implications of their work, ensuring that developments benefit local populations. Leaders should involve community stakeholders in the planning process, addressing concerns and prioritizing local needs. This approach fosters trust and ensures that projects are aligned with the community's vision for growth.

The concept of regenerative design takes sustainability a step further by aiming to restore and enhance the natural environment. This holistic approach considers the lifecycle of materials and the ecological impact of construction. By designing buildings that contribute positively to their surroundings, such as integrating green spaces or improving local ecosystems, construction firms can be at the forefront of a movement that seeks to create harmony between the built environment and nature.

The intersection of technology, leadership, and sustainability creates a powerful synergy that can drive significant advancements in the construction industry. As we explore their interdependencies, it becomes evident that they are not standalone elements; rather, they work together to create a cohesive foundation for modern construction practices like IPD and smart cities.

A notable approach that exemplifies this synergy is Integrated Project Delivery (IPD). IPD emphasizes collaboration among all stakeholders, leveraging technology and effective leadership to ensure that sustainability is at the forefront of decision-making. By fostering a team-oriented environment, IPD enhances communication and alignment, ultimately leading to projects that are completed more efficiently, on budget, and with a focus on reducing environmental impact.

The future of urban development is increasingly leaning towards the concept of smart cities, where technology, sustainability, and leadership converge to create livable, efficient spaces. Construction firms play a crucial role in building the infrastructure for smart cities, incorporating renewable energy sources, efficient public transportation, and advanced communication technologies. Leaders in construction must advocate for sustainable practices while utilizing technology to address the challenges of urbanization.

As sustainability becomes a priority, regulatory bodies and industry organizations are establishing standards that prioritize environmentally responsible practices. Leaders in the construction industry must stay informed and compliant with these regulations. By championing sustainability and technology adoption, construction firms can shape industry standards and demonstrate their commitment to a sustainable future.

As we navigate the complexities of modern construction, it's evident that technology, leadership, and sustainability are the foundational pillars for progress. Embracing these elements allows the industry to redefine traditional practices and create efficient, safe, and environmentally responsible structures.

Furthermore, to build a better future, construction professionals must commit to continuous learning and adaptation, ensuring that they remain at the forefront of innovation. By working together—leveraging technology, fostering effective leadership, and prioritizing sustainable practices—we can reshape the construction industry for the better and create a built environment that serves both people and the planet.

The journey towards modernization is just beginning, but with a strong foundation based on these essential pillars, the construction industry stands poised for a brighter, more sustainable future. Data-driven decision-making is the need of the hour.

AI-Powered paving with predictive maintenance techniques and autonomous machines has revolutionised the construction field.

Artificial Intelligence (AI) and machine learning are revolutionizing various industries, and the construction and paving sectors are no exception. The application of these advanced technologies in analyzing data concerning traffic patterns, weather conditions, and material performance can significantly enhance the efficiency and effectiveness of asphalt paving projects.

To begin with, analyzing traffic patterns is crucial for understanding how different factors influence road conditions and paving durability. Traffic data can provide insights into peak usage times, types of vehicles, and traffic loads. AI algorithms can process massive datasets from various sources, such as traffic cameras, sensors, and GPS data from vehicles, to create predictive models that estimate the effects of traffic on asphalt wear over time. By understanding these patterns, engineers can choose paving materials and techniques that are better suited to withstand the specific demands of different roadways.

For instance, roads that experience heavy truck traffic may require more robust paving materials or techniques than those used on residential streets. Machine learning can also facilitate the continuous monitoring of these roads, allowing for timely interventions and repairs before major issues arise. By leveraging real-time data, maintenance schedules can be optimized based on actual road conditions rather than predetermined timelines, leading to more efficient use of resources and extended pavement life.

In addition to traffic data, weather conditions play a significant role in the performance of asphalt pavements. Temperature fluctuations, rainfall, and humidity levels can all impact how asphalt materials behave over time. AI can analyze historical weather data along with current conditions to create models predicting when and how these environmental factors will affect paving materials. For example, AI algorithms can recommend adjustments in the paving schedule or modifications to the asphalt mix to ensure better performance and longevity of the pavement if a particular region experiences excessive rainfall.

Moreover, AI-driven predictive analytics can help in understanding the long-term effects of various weather conditions on pavement degradation. By analyzing data on how pavement materials react under different environmental scenarios, engineers can make informed decisions about which materials to utilize for specific weather patterns, thus enhancing the durability and performance of the pavement.

Material performance is another area where AI and machine learning are making significant advancements. With the development of new asphalt mixtures and paving techniques, it is essential to assess their performance under various conditions accurately. Machine learning algorithms can evaluate data from laboratory tests, field trials, and historical performance records to identify trends and correlations that might not be evident through traditional analysis methods. This approach allows for better prediction of how new materials will perform in real-world conditions, which can significantly influence material selection for paving projects.

For instance, machine learning can help identify the best asphalt mix for high-traffic roads in areas prone to extreme temperature fluctuations. By optimizing the composition of asphalt mixtures based on the predicted performance in given conditions, project managers can improve the overall quality and lifespan of the pavement while also reducing costs associated with premature failures and repairs.

The integration of AI in paving is not limited to the analysis of traffic, weather, and material performance; it also extends to the planning and visualization stages of asphalt paving projects through Building Information Modeling (BIM). BIM is a digital representation of the physical and functional characteristics of a project, and its use in asphalt paving is becoming increasingly prominent.

BIM allows for enhanced project planning by enabling engineers and project managers to visualize the entire paving process in a virtual environment. This 3D modeling tool helps stakeholders understand the project better, resulting in improved communication and collaboration among team members. By incorporating data from various sources, including traffic analysis and weather forecasts, BIM can simulate different scenarios and provide insights into the potential challenges that may arise during the project lifecycle.

One of the major benefits of using BIM in asphalt paving projects is the ability to manage costs effectively. With detailed models, project managers can estimate material quantities, labor costs, and equipment needs more accurately. By identifying potential cost overruns early in the process, teams can make adjustments to stay within budget, ensuring the project is completed successfully and economically.

BIM also facilitates better decision-making by providing a comprehensive view of the project. It allows project teams to assess various paving designs and materials side by side, evaluating their impacts on both performance and costs. Furthermore, the integration of AI into BIM can enhance predictive modeling capabilities, enabling teams to forecast better how different designs will perform over time under varying conditions.

In summary, the application of AI and machine learning in analyzing data related to traffic patterns, weather conditions, and material performance is transforming the asphalt paving industry. By leveraging these technologies, project managers can optimize paving techniques, improve material selection, and refine maintenance schedules. Additionally, the integration of BIM provides enhanced planning, visualization, and cost management capabilities, further enhancing the efficiency and effectiveness of paving projects.

Predictive maintenance, sensor-enhanced roads, and data-driven bitumen upgrades are innovative concepts gaining traction in the field of infrastructure management and transportation. Predictive maintenance involves using data analytics and sensor technologies to anticipate when maintenance should be performed, thereby reducing costs and downtime. By monitoring the condition of assets in real-time, organizations can proactively address issues before they escalate, improving safety and reliability.

Sensor-enhanced roads further support this initiative by embedding sensors into roadways, allowing for the collection of data on traffic patterns, weather conditions, and road wear. This data provides valuable insights that can inform infrastructure decisions, optimize traffic flow, and enhance driver safety.

Data-driven bitumen upgrades represent advancements in material science, where the performance and durability of road surfaces are improved through the use of advanced materials rooted in data analysis. These enhancements not only extend the lifespan of roads but also contribute to more sustainable construction practices.

While these concepts may initially appear theoretical, they are being implemented in various projects globally, leading to smarter, safer, and more efficient transportation systems that are better equipped to meet the demands of modern society. This practical application signifies a shift towards data-driven decision-making in infrastructure development and maintenance. Some notable examples are the European Smart Highways Initiative, Japan's IoT-Enabled Pavements, United States Interstate Upgrades, Smart Road Corridors in China, and Australia's Remote Highway Monitoring.

Numerous countries within the European Union are currently engaged in innovative initiatives involving the integration of embedded sensors in highways, which provide continuous data regarding traffic flow and structural integrity. Collaborative efforts between government agencies and private technology firms aim to predict potential areas of degradation, thereby minimizing the need for emergency closures.

Roadways are outfitted with high-sensitivity strain gauges capable of detecting micro-fractures caused by seismic activity in regions prone to earthquakes. By employing artificial intelligence analytics, officials can swiftly prioritize repairs along critical evacuation routes.

Transportation departments in states such as California and Texas are piloting predictive maintenance technologies that take into account factors such as extreme temperatures and heavy truck usage. The performance of polymer-modified asphalt is being evaluated in comparison to conventional mixtures, with sensor data playing a pivotal role in guiding future material selections.

China is at the forefront of adopting advanced infrastructure technologies, with certain provinces leading the way in the implementation of solar-powered roads that incorporate sensors and dynamic road markings. Artificial intelligence algorithms analyze the data to enable real-time toll pricing and route guidance.

In more remote and sparsely populated regions, extensive stretches of highway are monitored using satellites and drones that relay data to machine learning models. This methodology assists local authorities in deploying maintenance crews to isolated areas before road conditions deteriorate further.

These varied projects underscore that the transition to smart infrastructure transcends mere trends; it represents a significant evolution in constructing, monitoring, and maintaining roadways. This transformation capitalizes on a combination of advanced materials, AI-driven insights, and sensor technology.

As the industry increasingly adopts digital innovations, the integration of artificial intelligence (AI), machine learning, and Building Information Modeling (BIM) is set to foster a more intelligent and sustainable future for asphalt paving. The ability to comprehend and leverage the extensive amounts of data at their disposal allows stakeholders to make well-informed decisions. This approach will lead to the development of roads that exhibit enhanced performance, lower maintenance costs, and, ultimately, safer and more efficient transportation networks for all users.

Urban planners, in collaboration with civil engineers and policymakers, will gain considerable insights from these advancements. The utilization of AI and machine learning, in conjunction with BIM, enables the analysis of historical and real-time data, which can inform better design and construction practices. This synergy not only enhances the quality and durability of road networks but also promotes cost-effectiveness in their development and upkeep.

Moreover, the integration of these technologies is expected to contribute significantly to the optimization of resource allocation and project management, ensuring that infrastructure projects adhere to budgets and timelines. The proactive approach fostered by data-driven insights facilitates the identification of potential issues before they arise, thereby minimizing disruptions during construction and service.

Ultimately, the collaboration between technology and engineering expertise is essential for creating resilient transportation systems that meet the demands of urbanization and environmental sustainability. By embracing this technological evolution, stakeholders can contribute to a future that prioritizes the integrity of road networks, benefiting all community members who rely on these vital infrastructures for mobility and connectivity.

CHAPTER 6

THE GREEN ROAD AHEAD: SUSTAINABLE INNOVATION IN CONSTRUCTION

Across various regions, within and beyond, self-healing eco-friendly roads are being implemented, showcasing their potential through global examples and case studies.

In the Netherlands, researchers have incorporated steel fibers into asphalt to facilitate self-healing through induction heating, thereby reducing the frequency of maintenance. This technique not only prolongs pavement life but also diminishes environmental impact by minimizing raw material consumption.

Similarly, in the United Kingdom, microcapsule technology is utilized. These capsules release healing agents when cracks occur, maintaining structural integrity and optimizing lifecycle costs.

In Australia, the focus shifts to using recycled materials in self-healing road construction. Engineers have successfully incorporated waste polymers and by-products, enhancing road durability while promoting sustainability.

Additionally, in Sweden, roads are being designed with embedded sensors that monitor stress and predict maintenance needs, ensuring timely interventions that prevent further degradation.

Each of the above examples underscores the importance of integrating novel materials and technologies to enhance road longevity and ecological viability.

The intersection of environmentally friendly practices and technological innovation is increasingly vital in our quest for a sustainable future. As the effects of climate change become more pronounced, integrating green practices into technology is essential to reduce our ecological footprint. Technological advancements not only present innovative solutions but also open new avenues for enhancing sustainability across various sectors.

For instance, renewable energy technologies, such as solar panels and wind turbines, have evolved significantly. Innovations in energy storage, like advanced batteries and grid integration techniques, allow for the efficient utilization of renewable resources, minimizing dependence on fossil fuels. Additionally, smart grid technology enables better management of energy distribution, leading to reduced energy waste and improved efficiency.

Moreover, the rise of the circular economy, bolstered by innovative technologies, promotes recycling and sustainable production processes. Companies are now leveraging artificial intelligence and machine learning to optimize resource use and minimize waste in manufacturing. These technologies can analyze patterns and improve processes, leading to smarter resource management.

In agriculture, precision farming technologies help reduce water usage and chemical inputs while increasing crop yields. Drones and IoT devices allow farmers to monitor their resources in real time, ensuring that efforts are made only where necessary.

As consumers, adopting environmentally conscious habits can drive demand for such innovations. By supporting businesses that prioritize sustainability, individuals can contribute to a larger movement towards environmental stewardship. Ultimately, the fusion of eco-friendly practices with cutting-edge technology paves the way for resilient ecosystems, healthier communities, and the promise of a sustainable future for generations to come.

Sustainability in road construction has become increasingly vital in addressing environmental concerns and promoting long-term economic viability. As infrastructure projects expand, the environmental impact of traditional construction methods becomes more apparent, driving the need for greener alternatives. Sustainable road construction involves using eco-friendly materials, minimizing waste, and reducing emissions during the building process. Techniques such as recycling materials, employing energy-efficient machinery, and designing roads that enhance natural drainage systems contribute to mitigating harm to ecosystems. Furthermore, sustainable practices can lead to cost savings over time through reduced maintenance needs and energy consumption. Adopting sustainable approaches not only benefits the environment but also fosters community well-being by creating safer, more resilient infrastructure. As urban areas continue to grow, integrating sustainability into road construction will be essential to ensure that development meets current needs without compromising the ability of future generations to meet theirs.

It is a fact that road construction plays a critical role in economic development, providing essential infrastructure that facilitates transportation, trade, and communication. However, traditional road construction practices often lead to significant environmental degradation, including habitat loss, pollution, and high carbon emissions. As the world increasingly acknowledges the need for sustainability, the road construction industry is gradually transforming, embracing sustainable innovations that address environmental concerns and enhance efficiency and longevity. In this chapter, I will delve into the various sustainable practices and innovations currently shaping the future of road construction.

First, it is important to understand the idea of sustainable road construction. Sustainable road construction involves the integration of environmental friendly practices throughout the lifecycle of roadways—from planning and design to construction, maintenance, and decommissioning. The aim is to minimize negative environmental impacts while maximizing economic and social benefits. Key principles include reducing resource consumption, minimizing waste, avoiding pollution, and enhancing infrastructure resilience.

One of the most significant areas of sustainable innovation in road construction is the development of alternative materials. Traditional asphalt and concrete are resource-intensive and have a considerable carbon footprint. Innovations in materials science are leading to the use of recycled materials, green concrete, and bio-based materials.

Incorporating recycled asphalt pavement (RAP) and reclaimed concrete aggregates (RCA) reduces the need for virgin materials like green concrete and bio-based materials, and lowers overall construction costs.

Green concrete is an eco-friendly concrete that utilizes industrial by-products, like fly ash and slag, to replace a portion of Portland cement, thereby reducing carbon emissions generated during production. SCMs usage reduces the amount of Portland cement required, lowering concrete's carbon footprint. Its usage also reduces the amount of waste material sent to landfills, making it a more sustainable and environmentally friendly option. Due to the use of SCMs, green concrete has better long-term strength and durability compared to traditional concrete. This makes it an excellent choice for high-traffic areas such as roads, bridges, and airports, as it can withstand heavy loads and frequent use.

In India, the Delhi Metro Rail Corporation (DMRC) used a mix of 50% fly ash and 50% slag in the construction of its metro stations, reducing the amount of Portland cement used by 40%. The use of green concrete in the construction of the DMRC stations resulted in a significant reduction in CO2 emissions and waste materials.

In Vancouver, Canada, a company called Nexii Building Solutions has developed a green concrete alternative called Nexiite. Nexiite is made using 90% less water and 70% less cement than traditional concrete, making it a more sustainable and eco-friendly option for building homes and other structures.

Bio-based materials are innovations that have led to the exploration of bio-asphalt from natural sources like plant oils, hemp shives, sunflower, straw, cellulose, fungus, and resins. They provide sustainable alternatives to traditional petroleum-based products.

Eco-friendly construction techniques are an important aspect of sustainable development. The construction processes themselves are being transformed through eco-friendly methods. Some of these innovations include the Warm Mix Asphalt (WMA) technique, which allows asphalt to be produced and laid at lower temperatures, reducing energy consumption and emissions during the mixing and application processes.

Stabilized Earth Structures utilize local soil stabilization techniques that can reduce the need for imported materials and minimize construction waste. Modular Construction techniques involve prefabricated components that can enhance construction efficiency and reduce on-site waste, minimizing disturbance to surrounding ecosystems. The next important factor is Sustainable Design Practices. Sustainable innovation starts with thoughtful planning and design. Road engineers can implement various design strategies to enhance sustainability. For instance, Low-Impact Development (LID) involves incorporating permeable surfaces and green stormwater management techniques. It can reduce runoff and improve water quality, preserving local ecosystems.

Biodiversity Considerations in designing roads to accommodate wildlife corridors help maintain animal habitats and mobility, thus minimizing ecological fragmentation.

Smart Road Design involves integrating technology in road design to optimize traffic flow, reduce congestion and minimize vehicle emissions. Adaptive traffic signals, for instance, can respond to real-time traffic conditions to improve efficiency.

Technology plays a pivotal role in revolutionizing road construction. Some of the groundbreaking technologies include drones and aerial surveys. Drones can conduct surveys and assessments of construction sites more efficiently than traditional methods, providing accurate data that improves planning and reduces costs.

3D Printing allows for rapid and efficient construction of road components, reducing material waste and enabling greater precision in construction. BIM enables enhanced visualization and simulation of road projects, allowing planners and engineers to identify potential problems early in the design phase and optimize resource use.

Energy Efficiency in Construction is also of extreme importance. Energy consumption during the construction phase can significantly impact sustainability. Innovations in energy efficiency include renewable energy sources. Utilizing solar and wind energy on construction sites for operations can significantly reduce carbon footprints. Portable solar panels can provide power for equipment and site lighting.The adoption of electric and hybrid vehicles limits emissions generated on-site, contributing to a greener construction process.

My research paper on 'Road to Net Zero: A Carbon Emissions Calculator for Pavement Materials and Processes' talks about the tool developed to curb carbon emissions. The asphalt pavement industry is one of the significant contributors to global greenhouse gas (GHG) emissions, largely due to energy-intensive production processes and material transportation. To address the environmental impact, we proposed a conceptual framework for developing a carbon calculator tailored to the asphalt industry. This tool is being designed to provide lifecycle-based carbon accounting, enabling stakeholders to measure emissions across all phases of pavement projects. Additionally, the tool integrates a carbon credit aggregation mechanism, empowering users to offset residual emissions through bulk purchases in verified carbon markets. By bridging carbon accounting with offset procurement, this innovative tool supports sustainable practices and incentivizes decarbonization in infrastructure development. This paper outlined the methodology, potential impact, and future directions for this transformative initiative.

Zero carbon emission is aided by maintaining a lifecycle assessment report. Lifelong Maintenance and Lifecycle Assessment is a sustainable approach that does not end with the construction of roads. Lifelong maintenance strategies and lifecycle assessments are critical for ensuring long-term sustainability, including regular monitoring. Using technology for real-time monitoring can help detect road deterioration early, enabling proactive maintenance that extends the life of the infrastructure.

Life Cycle Cost Analysis (LCCA) involves evaluating the economic implications of road construction options over their entire lifespan, ensuring that choices made during the planning stage align with sustainability goals. Policy and Regulation Influence hugely impacts sustainability. Government policy and regulations are vital in encouraging sustainable innovation in road construction. Incentives for sustainable practices encourage consumers and stakeholders to take sustainable practices seriously. Governments can provide tax breaks, grants, or low-interest loans for projects that incorporate sustainable technologies or materials. Stricter regulations surrounding emissions and environmental impact assessments can promote more sustainable construction methods and practices.

Another concept is Public-Private Partnerships (PPPs) that we need to know about. These are collaborative projects that can bring investment, resources, and innovation to sustainable road construction while balancing economic and environmental objectives.

Community Engagement and Social Responsibility are social factors influencing the sustainability aspect of road construction. The social dimensions of sustainability are equally important. Engaging with local communities during the planning and construction phases fosters goodwill and ensures that projects meet the needs of residents. Strategies include early communication or consultation with the community regarding project impacts and benefits. It is essential. This inclusion is crucial for gathering local knowledge, values, and concerns.

Sustainable road construction often demands new skills. Providing training for workers in new technology and sustainable practices can create employment opportunities and foster economic growth.

This seems doable, but there are challenges and future directions that cannot be ignored. While the promise of sustainable innovations in road construction is significant, challenges remain. These include initial costs. Many sustainable materials and technologies may have higher upfront costs, necessitating financial incentives or support to encourage adoption.

Established practices within the industry can hinder the swift incorporation of innovative approaches, emphasizing the need for education and awareness. Resistance to change should be avoided. Keeping pace with rapid technological advancements requires ongoing education and investment in research and development.

Sustainable innovation in road construction is not just a necessity; it presents an opportunity for creating resilient infrastructure that benefits the environment, the economy, and society at large. Through the adoption of alternative materials, eco-friendly construction techniques, advanced technologies, and community engagement, the road construction industry can pave the way toward a sustainable future. As the sector continues to evolve, collaboration between governments, industry stakeholders, and communities will be essential for overcoming challenges and ensuring a greener, more sustainable transportation infrastructure for generations to come.

Some other aspects of road construction provide a framework for sustainable development, ensuring that human activities can coexist with the planet's ecosystems. The discourse surrounding sustainability has gained momentum in recent years as the impacts of climate change and resource depletion become increasingly evident. Among the most prominent concepts emerging in this dialogue are circular construction, renewable energy, and carbon-neutral projects. I will talk about each of them briefly.

Circular construction is a paradigm shift from the traditional linear model of "take-make-dispose." Instead of following a one-way path, circular construction emphasizes the importance of reusing, recycling, and repurposing materials throughout the entire lifecycle of a building or structure. This approach minimizes waste and reduces the consumption of finite resources.

At its core, circular construction integrates principles of ecology into the built environment. This includes designing buildings for disassembly, where materials can be easily retrieved and reused at the end of a building's life, thus extending the life cycle of both materials and the buildings themselves. For example, modular construction techniques enable the creation of structures with components that can be replaced or upgraded without demolishing the entire framework.

Another critical aspect of circular construction is the use of sustainable materials. These are often derived from renewable resources or are designed to be recycled or composted. Innovators are exploring options such as bio-based materials, reclaimed timber, and even materials made from industrial by-products. This not only lessens the environmental impact but can also drive economic growth by fostering new industries focused on sustainable material production and waste management.

Moreover, circular construction takes into account the entire ecosystem. By utilizing local materials and labor, it reduces the carbon footprint associated with transportation while bolstering local economies. Sustainable urban planning that considers the social, economic, and environmental aspects can lead to more resilient communities.

The transition to renewable energy is a cornerstone of sustainable development. It encompasses energy sources that are naturally replenished, such as sunlight, wind, rain, tide, waves, geothermal heat, and biomass. The most significant advantage of renewable energy is its potential to reduce greenhouse gas emissions significantly when used instead of fossil fuels.

Solar and wind energy have seen dramatic growth in recent years, with technological advancements leading to increased efficiency and reduced costs. For instance, the cost of solar photovoltaic (PV) cells has plummeted over the last decade, making solar power more accessible to consumers and businesses alike. On a larger scale, wind farms have become a critical part of energy generation in many countries, showcasing how strategic investments in renewable infrastructure can lead to substantial energy independence and stabilization of energy prices.

Importantly, renewable energy contributes to energy security. By diversifying energy sources and relying less on fossil fuel imports, nations can mitigate the risks associated with volatile oil markets and geopolitical tensions. Moreover, decentralized renewable energy systems—such as community solar projects—empower local communities by providing them with greater control over their energy sources.

However, transitioning to renewable energy also presents challenges. The intermittent nature of some renewable sources, like solar and wind, necessitates the development of robust energy storage solutions and grid management technologies. Innovations in battery storage and smart grids are crucial for maximizing the potential of renewable energy.

Carbon neutrality—or net-zero carbon emissions—refers to the balance between the amount of greenhouse gas emissions produced and the amount removed from the atmosphere. Achieving carbon neutrality is essential for mitigating climate change and involves reducing emissions wherever possible and investing in solutions that absorb or offset the remaining emissions.

Carbon-neutral projects are increasingly gaining traction across various sectors, including real estate, manufacturing, and transportation. Buildings designed to be energy-efficient, utilize renewable energy, and incorporate sustainable materials contribute significantly to carbon neutrality. For instance, passive house designs that maximize natural light and insulation can drastically reduce energy consumption.

In industry, manufacturers are innovating processes to minimize waste and optimize energy use. The implementation of energy management systems can optimize production schedules around times of renewable energy generation, further reducing reliance on fossil fuels. Additionally, technologies such as carbon capture and storage (CCS) are being explored to capture carbon emissions from industrial processes before they enter the atmosphere.

Transportation is another critical area for achieving carbon neutrality. The shift towards electric vehicles (EVs), supported by an expanding network of renewable energy-powered charging stations, represents a significant step toward reducing emissions in this sector. Furthermore, public transportation innovations, such as electric buses and the promotion of cycling and walking infrastructure, contribute to reducing the overall carbon footprint of urban environments.

I will discuss electricity-run vehicles and the required infrastructure in another chapter.

The intersection of circular construction, renewable energy, and carbon-neutral projects creates a holistic approach to sustainability. Each concept reinforces the others: circular construction can enhance the sustainability of renewable energy infrastructure, while renewable energy significantly contributes to the carbon-neutral goals of projects.

Adopting these principles is not merely about compliance or corporate social responsibility; it is a fundamental shift required for the planet's survival and future prosperity. As we continue to face the harsh realities of climate change and resource scarcity, the integration of these concepts into our urban fabric, industries, and lifestyles will determine our ability to thrive sustainably.

In conclusion, fostering collaboration among governments, businesses, and communities will best promote the circular economy, the transition to renewable energy, and the pursuit of carbon neutrality. By embracing these concepts, society can build a more resilient, equitable, and sustainable future that respects both people and the planet.

The discussion on innovations is incomplete without talking about RTGES. Asphalt pavements are ubiquitous in urban areas, serving as critical infrastructure for roadways, parking lots, and other surfaces. However, these surfaces absorb and retain heat from solar radiation, contributing to the urban heat island effect and increasing energy consumption for cooling. The concept of harnessing this excess heat for energy generation is a compelling solution to address both energy

needs and environmental concerns. The development of Thermoelectric Generator Systems (RTEGS) offers a transformative opportunity to convert heat from asphalt pavements into usable electricity.

Thermoelectric generators (TEGs) operate based on the Seebeck effect, where a temperature difference across a thermoelectric material generates an electric voltage. When one side of a thermoelectric material is heated (in this case, by the warmer asphalt) while the other side remains cooler (potentially aided by ambient air or cooling features), a voltage is created. If connected to a circuit, this voltage can drive electrical current and be harnessed for various applications, such as powering lights, sensors, or even contributing to the grid.

There are numerous advantages of using asphalt for energy harvesting. To begin with, the use of asphalt pavements leverages renewable solar energy, making TEGs a sustainable option. Given that asphalt is widely used, the energy harvested can be significant, especially in urban environments with extensive road networks.

Implementing thermoelectric generators can help reduce surface temperatures of asphalt, thereby alleviating the urban heat island effect. By converting heat to electricity, road surfaces can remain cooler, further benefiting in terms of energy consumption and thermal comfort.

They function dually. RTEGS can provide energy while simultaneously serving their primary purpose as a roadway. This dual functionality allows the integration of energy harvesting technologies into existing infrastructure without requiring significant changes.

While the concept of using RTEGS is promising, several challenges need to be addressed for successful implementation, like material efficiency, durability, cost-effectiveness, and energy storage and management. The efficiency of thermoelectric materials is crucial for energy conversion. Current materials may not allow for optimal voltage generation under the temperatures present in asphalt pavements. Research is needed to discover or develop new materials with higher thermoelectric efficiency. Pavements experience heavy vehicular loads and environmental stress. Any installed thermoelectric systems must be robust and durable enough to withstand wear and tear, temperature fluctuations, and exposure to the elements. The installation and maintenance costs must be justified by the energy produced. A comprehensive economic analysis will help gauge feasibility and identify potential funding or incentives for such projects. Energy harvested during peak temperature periods might need to be stored for use during lower energy demand periods. Developing efficient energy storage solutions integrated with RTEGS will be necessary to maximize the benefits.

To effectively develop RTEGS for energy harvesting from asphalt pavements, several strategic approaches can be employed. For instance, pilot programs, integration with smart technologies, collaboration with stakeholders, community engagement, and environmental assessments.

Implementing small-scale pilot projects in various urban settings will help assess the feasibility, energy output, and adaptability of RTEGS. Pilot programs will provide valuable data and refine system parameters for larger-scale deployment. Combining RTEGS with smart grid technologies can enhance their functionality. Real-time monitoring of energy generation, consumption, and pavement conditions can optimize performance and facilitate better energy management. Engaging city planners, transportation agencies, and academic institutions will promote knowledge sharing and innovation. Exploring public-private partnerships can also aid funding and technical expertise.

Educating communities about the benefits of energy harvesting from pavements may foster public support for these initiatives. Local awareness campaigns can encourage public involvement and interest in sustainable practices. Conducting comprehensive assessments of potential environmental impacts will help in understanding how RTEGS can contribute to local sustainability goals. Studies on the effect of reduced surface temperatures on local ecosystems and air quality will further justify the installations.

Countries like the US, Japan and China are investing hugely in these generators. Other countries are also following suit. The future of RTEGS for energy harvesting from asphalt pavements holds great promise. As technology evolves, advances in materials science and engineering may lead to the development of more efficient and economically viable thermoelectric systems. Furthermore, as global energy demands rise, innovative solutions like RTEGS could play a pivotal role in diversifying renewable energy sources and promoting sustainable urban development.

The development of thermoelectric generator systems for energy harvesting from asphalt pavements presents an exciting intersection of renewable energy technology and urban infrastructure. By converting waste heat into electricity, these systems not only help mitigate urban heat effects but also contribute to more sustainable energy practices. Through collaborative research, pilot projects, and effective stakeholder engagement, RTEGS can revolutionize how cities harness energy from existing infrastructure, ultimately leading to a more sustainable future.

In recent years, the construction industry has increasingly focused on sustainable practices and innovative technologies to enhance the performance and longevity of road surfaces by using advanced and eco-friendly solutions in asphalt technology. Among these advancements, two notable concepts have emerged: self-healing asphalt and permeable asphalt. These solutions are not only designed to improve the functionality of pavements but also to address pressing environmental concerns. In the Netherlands, the HEALROAD project demonstrated the effectiveness of induction heating in extending the lifespan of asphalt pavements. Trials with microcapsule technology have shown improved durability and reduced maintenance in high-traffic areas in France. In China, Bitumen rejuvenation projects have highlighted the feasibility of large-scale applications, showing significant reductions in maintenance needs and costs.

Self-healing asphalt represents a groundbreaking approach to road maintenance and durability. The essence of this technology lies in its ability to autonomously repair small cracks and fissures that develop in pavement over time. Traditional asphalt surfaces are susceptible to wear and tear due to environmental factors, traffic loads, and temperature fluctuations. These cracks, if left unaddressed, can lead to significant degradation, requiring costly maintenance and repairs. In the Netherlands, several highways have been constructed using induction-heated self-healing asphalt, demonstrating significant improvements in pavement performance and longevity.

The self-healing properties of asphalt can be achieved through various innovative methods. One prominent technique involves incorporating special polymers or microcapsules filled with a healing agent into the asphalt mix. When the asphalt experiences stress and cracks form, these microcapsules rupture, releasing the healing agent into the damaged area. This agent then reacts with the asphalt to form a solid, cohesive bond, effectively sealing the crack and mitigating further deterioration.

Another approach involves the use of shape memory polymers, which can undergo deformation and revert to their original shape under specific conditions, effectively sealing cracks without additional intervention. Researchers are also exploring the use of bio-based materials that enhance the self-healing capabilities, making the asphalt not only more resilient but also more environmentally friendly.

The benefits of self-healing asphalt are multifold. By minimizing the need for frequent repairs, self-healing asphalt can significantly reduce maintenance costs for municipalities and private entities alike. Self-healing properties can prolong the lifespan of pavement, leading to lower lifecycle costs and reduced resource consumption. With fewer repairs required, the carbon footprint associated with the production and transportation of materials for road repairs is diminished. As cracks can lead to potholes and uneven surfaces, the use of self-healing asphalt can enhance road safety and overall performance for vehicles. Researchers continue to study and refine self-healing mechanisms, paving the way for more adaptable solutions tailored to different environmental conditions and traffic demands.

Permeable asphalt is another innovative solution that addresses not just the structural integrity of pavements, but also the environmental challenges associated with stormwater runoff. Traditional asphalt surfaces tend to be impermeable, preventing water absorption and contributing to flooding in urban areas. Permeable asphalt, on the other hand, allows for water to pass through the surface, promoting sustainable stormwater management.

Permeable asphalt is engineered with a specific composition that includes larger aggregates and a reduced amount of fine materials, creating voids that permit water to flow through. This design facilitates the natural infiltration of rainwater into the ground below, rather than allowing it to accumulate on the surface or flow directly into storm drains.

To further enhance performance, permeable asphalt systems often incorporate layers of aggregate beneath the surface that act as reservoirs, temporarily holding water before it drains into the soil. This approach not only assists with flood mitigation but can also help in recharging local groundwater supplies.

Stormwater management, water quality improvement, urban heat island mitigation, biodiversity support, and sustainability are some significant advantages of permeable asphalt.

By allowing water to permeate through the pavement, permeable asphalt reduces surface runoff and the risks of flooding, especially in urban areas subject to heavy rainfall. The filtration provided by the layers of aggregates helps in capturing pollutants, sediments, and debris, thus improving the quality of water re-entering the natural ecosystem. Permeable surfaces can reduce the urban heat island effect, as they generally remain cooler than traditional impervious asphalt, promoting cooler urban environments. Effective stormwater management can help maintain local ecosystems, supporting flora and fauna that may depend on a balanced water table. The implementation of permeable asphalt contributes to sustainable urban development practices by addressing environmental concerns related to water management.

Despite the promising benefits of self-healing and permeable asphalt technologies, several challenges must be addressed for their broader adoption, which are cost and investment, material performance, maintenance and longevity, regulatory and standardization issues, and finally, public awareness and acceptance.

The initial costs associated with developing and implementing these advanced technologies can be significant. Governments and municipalities must weigh the long-term benefits against the upfront expenses. Ongoing research and testing are essential to ensure that the self-healing and permeable asphalt can withstand high traffic loads, extreme weather conditions, and other stressors associated with roadways. While self-healing asphalt is designed to reduce the need for repairs, it is crucial to establish protocols for monitoring and maintaining its effectiveness over time. Similarly, the long-term performance of permeable asphalt must be evaluated to ensure its continued efficacy in managing stormwater. The adoption of new materials and technologies often requires regulatory approval and the establishment of standards. Collaboration between researchers, engineers, and policymakers is essential to facilitate the integration of self-healing and permeable asphalt into mainstream practices. As with any new technology, educating the public and stakeholders about the benefits and functionality of these advanced asphalt solutions is key to ensuring successful implementation.

The exploration of self-healing and permeable asphalt technologies highlights an exciting intersection of innovation and sustainability within the construction industry. As urban areas continue to grapple with challenges such as infrastructure maintenance, flooding, and environmental degradation, these advanced solutions offer compelling options that can enhance road durability while promoting eco-friendly practices.

By investing in research and development and by fostering collaboration among stakeholders, we can pave the way for a future where roads are not just functional but also contribute positively to the environment. The potential of self-healing and permeable asphalt illustrates a forward-thinking approach to infrastructure that aligns with the needs of modern society and the health of our planet. As these technologies continue to develop, we may see a significant shift in how we design, build, and maintain our roadways, ultimately leading to safer and more sustainable transportation systems.

CHAPTER 7

CONSTRUCTION 4.0: THE REVOLUTION OF INNOVATION AND LEADERSHIP

Construction 4.0 is the digitization of Industrial 4.0. We can also say that it is the realization of the digital transformation in the field, delivering real-time decision-making, enhanced productivity, and flexibility.

Though there is no global agreement on what exactly Construction 4.0 means, some people think of it as applying the Industry 4.0 concept to construction, focusing on using technology for quick decision-making through constant connectivity. Others believe it's about finding a way to harmonize the new tech trends in construction, while for a few, it's a broader strategy to tackle the industry's current challenges beyond just technology.

Regardless of how you define it, the main shift brought by Construction 4.0 seems to be about better connecting the physical world with the digital one through constant connectivity. We already see this connection happening in construction with techniques like BIM. This technology allows detailed digital models of projects and even links the construction site to these models. However, we still rely on people to manage and maintain this connection. With Construction 4.0, more technologies are expected to take over some of those human roles, leading to a smoother blend of the real world and its digital counterpart.

Let me explain through an example. Great Wall Motor (GWM) is a Chinese automobile manufacturer headquartered in Baoding, Hebei, China, known for producing SUVs and pickup trucks. The Great Wall Motor (GWM) has its second-largest manufacturing plant outside China. Its smart car factory is in Thailand. GWM's smart factory uses intelligent robots to automate various manufacturing processes.

In India, GWM India R&D center is in Bangalore, focusing on software development for hybrid and battery-operated electric vehicles. It also develops artificial intelligence for self-driving cars.

Similarly, in the construction industry, bricklaying robots can lay bricks much faster than human workers, increasing overall construction speed and reducing project timelines. Autonomous vehicles and drones can automate material transport, surveying, and inspection tasks, saving time and improving efficiency.

The Semi-Automated Mason or SAM, a robot created by a New York-based company, is so efficient at brick-laying that it promises to increase productivity while reducing overall labor costs. Manufacturing efficiency has increased significantly due to robotics and technology, while the efficiency on construction sites has remained stagnant for the last 20 to 30 years. Construction Robotics created SAM to solve that problem. SAM requires a human partner to smooth over the work, but the heavy lifting is left to the bot. The robot can lay bricks at least three times faster than humans. Moreover, it never gets tired or makes mistakes.

The construction industry, often regarded as one of the most traditional and slow-to-adopt sectors, is undergoing a profound transformation. At the heart of this transformation is the Fourth Industrial Revolution (4IR), a period marked by the fusion of digital, physical, and biological technologies. Automation, digitalization, and smart systems play a pivotal role in accelerating this revolution, reshaping the way construction projects are planned, designed, built, and managed. These technologies are not just augmenting traditional methods but are redefining the construction landscape, driving efficiency, safety, and sustainability. In this chapter, I will discuss how these elements contributed to the Fourth Industrial Revolution in construction, their impact on the industry, and the challenges and opportunities they present.

To begin with, I can say, Industry 4.0 means smart manufacturing, and Construction 4.0 means smart construction. The key technologies driving this evolution include the Internet of Things (IoT), Digital Twins, additive manufacturing, cloud computing, Cyber-Physical Systems (CPS), and Building Information Modeling (BIM).

The significant potential of the Construction 4.0 transformation lies in the near-complete automation of the entire project life cycle. This automation incorporates digital twins at every stage, from planning to operation, encompassing design and construction. During the design phase, the enhanced utilization of BIM models as digital twins allows simulations that assess the physical product (such as a building), its constructability and profitability (4D, 5D), as well as its sustainability (for instance, energy analysis).

Within a Construction 4.0 environment, the management of data will assume a pivotal role and necessitate the emergence of new business models in the industry. Beyond the diverse array of data and information typically generated during the design and execution of construction projects, an extensive volume of data will arise from sensors and other CPS technologies. The scale, variety, and requisite speed of this data will demand the deployment of platforms designed to interpret, manage, and specifically connect with existing business systems. Consequently, the systems required will extend beyond mere Big Data or decision-making platforms; they must also serve as matchmaking platforms, delivering innovative services and products that link construction sites with smart factories, as well as enhancing smart design and utilization of connected smart buildings (or cities). Naturally, numerous challenges remain that must be addressed, including the establishment of global standards on a large scale.

Automation, digitalization, and smart systems are at the forefront of this change, offering solutions to long-standing problems while also providing new opportunities for innovation.

Automation refers to the use of technology to perform tasks that would traditionally require human labor. In construction, automation encompasses a wide range of applications, from the use of robots and drones for site surveying and inspections to the integration of automated machinery in building processes.

Robotics in construction is an emerging field that promises to revolutionize the industry. Robots can perform tasks such as bricklaying, welding, and even 3D printing of buildings. These robots work with precision and consistency, reducing the risk of human error and improving efficiency. Additionally, drones are being increasingly used for surveying, inspection, and monitoring of construction sites. Drones provide real-time aerial views of a site, enabling project managers to monitor progress, assess safety, and detect issues that might be difficult to identify from the ground. The use of drones and robots can significantly reduce the need for manual labor, leading to faster construction times and lower costs.

Construction equipment such as excavators, bulldozers, and cranes is becoming more automated, allowing for greater precision and control in operations. For example, autonomous bulldozers can perform earth-moving tasks without human intervention, and automated cranes can lift and position materials with high accuracy. These machines are equipped with sensors, GPS, and AI algorithms that enable them to operate safely and efficiently, reducing the risk of accidents and improving productivity on the job site. Autonomous construction vehicles can work around the clock, increasing productivity and reducing labor costs. Moreover, these vehicles can communicate with each other and with other systems on the construction site, ensuring that tasks are completed most efficiently and safely.

Artificial intelligence (AI) and machine learning (ML) are playing an increasingly important role in construction, helping to automate decision-making and improve project management. AI can be used to analyze large datasets, identify patterns, and make predictions, allowing construction managers to make more informed decisions. For example, AI algorithms can be used to optimize construction schedules by analyzing historical project data and identifying the most efficient sequence of tasks. AI can also help improve safety on construction sites by analyzing real-time data from cameras and sensors to detect hazards and alert workers to potential risks.

One of the most exciting developments in construction automation is the rise of 3D printing technology. 3D printing in construction allows for the creation of complex structures and components using a wide range of materials, including concrete, plastics, and metals. This technology has the potential to drastically reduce construction time and labor costs, as well as improve the sustainability of buildings by minimizing waste. For example, 3D-printed homes have already been built in several locations around the world, demonstrating the potential for automation to address housing shortages and provide affordable housing solutions.

Digitalization refers to the integration of digital technologies into every aspect of construction, from design to execution. The adoption of digital tools has been a game-changer for the construction industry, enabling more efficient workflows, better communication, and more accurate decision-making.

Building Information Modeling (BIM) is one of the most significant advancements in digitalization for the construction industry. BIM is a digital representation of the physical and functional characteristics of a building or infrastructure. It allows architects, engineers, and contractors to collaborate on a single platform, ensuring that all stakeholders have access to accurate, up-to-date information throughout the project lifecycle. BIM helps in visualizing the design, simulating construction processes, and detecting potential issues before they arise, reducing costly errors and delays. The use of BIM is also instrumental in improving the sustainability of construction projects by enabling more accurate material estimates and better resource management.

Cloud computing has made it easier for construction companies to store and share vast amounts of data in real-time. Cloud-based platforms enable project teams to access and update documents, drawings, and schedules from anywhere, improving collaboration and ensuring that everyone is on the same page. Data analytics, when combined with cloud computing, allows for the analysis of large datasets to gain insights into project performance, predict potential risks, and optimize processes. For instance, predictive analytics can help forecast delays, budget overruns, and other issues, enabling project managers to take proactive measures before problems escalate.

Smart Systems are hugely influencing construction projects. The Internet of Things (IoT) has introduced a new era of connected devices in construction. IoT-enabled sensors are used to monitor equipment, materials, and even worker health and safety. For example, sensors can track the condition of machinery in real-time, alerting operators to potential malfunctions before they cause costly breakdowns. Similarly, wearable devices such as smart helmets and vests can monitor workers' vital signs and detect hazards such as extreme heat or heavy lifting, enhancing safety on construction sites. IoT is also used to track materials and inventory, ensuring that the right supplies are available at the right time, reducing delays and waste.

Digital twin technology creates a virtual replica of a physical asset, such as a building or a piece of infrastructure, which can be used to simulate and monitor its performance throughout its lifecycle. By integrating real-time data from sensors, digital twins provide valuable insights into the operation of a building or infrastructure, enabling more effective maintenance, energy management, and performance optimization. For example, a digital twin of a building can help facility managers monitor energy consumption, track wear and tear on building systems, and predict when maintenance will be needed, leading to more efficient and cost-effective operations.

Smart buildings are designed to be highly efficient, adaptable, and responsive to the needs of occupants. They integrate IoT sensors, AI, and other technologies to monitor and control various building systems, such as lighting, heating, ventilation, and air conditioning (HVAC). For example, a smart building can automatically adjust its lighting and temperature based on occupancy, reducing energy consumption and improving comfort. Additionally, smart infrastructure, such as smart roads and bridges, can monitor their condition and alert maintenance teams when repairs are needed, extending the lifespan of these assets and reducing maintenance costs.

While the integration of automation, digitalization, and smart systems in construction offers numerous benefits, it also presents several challenges. One of the main challenges is the high initial cost of implementing these technologies, which can be a barrier for smaller construction companies. Additionally, there is a need for skilled workers who can operate and maintain advanced technologies, which requires investment in training and education.

Another challenge is the fragmentation of the construction industry. The construction ecosystem consists of various stakeholders, including architects, engineers, contractors, and suppliers, who often work in silos. This lack of coordination can hinder the successful implementation of digital tools and automation systems. To address this, the industry must embrace greater collaboration and standardization of technologies and processes.

Despite these challenges, the opportunities presented by automation, digitalization, and smart systems are immense. By embracing these technologies, the construction industry can become more efficient, sustainable, and resilient. The innovation potential is vast, from the creation of fully automated construction sites to the development of smart cities designed to be sustainable and adaptable to changing needs.

The discussion on automation is incomplete without Automated Paving Systems, which use GPS-controlled machines and Robotic

Systems for precise asphalt application. In recent years, the construction industry has remarkably advanced in automated paving technology. Automated paving systems, especially those using GPS-controlled paving machines and robotic systems, are revolutionizing asphalt application to roadways and other surfaces. These systems are improving precision, efficiency, and overall quality while also reducing costs, enhancing worker safety.

I will explore the development and implementation of GPS-controlled paving machines and robotic systems, focusing on how they contribute to more accurate and consistent asphalt application.

Historically, paving roads involved manual labor and less precise machinery. Traditional pavers were operated by skilled workers who relied on their experience to determine the correct thickness and consistency of the asphalt. While these methods were effective, they lacked the precision and speed that modern technology offers.

The introduction of automated paving systems began with the integration of GPS (Global Positioning System) technology. GPS technology, which allows accurate real-time location tracking, provided a major leap forward in construction automation. By coupling GPS with advanced control systems, paving machines were able to apply asphalt with a level of precision previously unattainable. The ability to control the machine's position with millimeter-level accuracy ensures that the asphalt is laid with exceptional precision.

The core function of a GPS-controlled paving machine is to ensure that the asphalt is laid evenly and consistently across the surface. The process begins with the creation of a digital map of the project site, which is typically created using satellite imagery or ground-based surveying equipment. This map serves as the baseline for the paving machine's path and provides information about the slopes, elevations, and curves of the road.

Once the map is created, the GPS on the paving machine can compare the machine's position with the digital map to ensure it is on the correct path. The control system adjusts the speed, direction, and application rate of the asphalt to match the design specifications in real-time. This automated feedback loop significantly reduces the chances of human error and ensures that the asphalt is laid evenly across the surface.

A real-world example of this is the use of Caterpillar AP555F Mobil-Trac Paver in large infrastructure projects. This paver uses GPS and other sophisticated technologies to control the paver's movement and material application, ensuring an even layer of asphalt. This capability improves the surface quality, reducing the need for post-paving corrections and providing long-lasting durability.

In addition to ensuring proper placement, GPS-controlled paving machines can also help optimize material usage. By precisely controlling the amount of asphalt applied at any given moment, these systems can reduce material waste and improve the efficiency of the paving process. This is particularly important in large-scale projects where even small inefficiencies can lead to significant cost increases.

For instance, during the paving of Interstate 5 in California, a GPS-guided paver was used to apply the asphalt with high precision. This system enabled the team to monitor and adjust the material spread across the pavement in real-time, preventing excess material from being applied. As a result, the project saved on asphalt costs and minimized the environmental impact of material overuse.

Volvo's HX2 autonomous hauler, used on construction sites, is capable of hauling materials to paving machines without requiring a human driver. This reduction in the need for manual labor helps contractors save on wages, as fewer workers are needed to perform essential tasks. The integration of autonomous vehicles and robotic systems can also minimize downtime between operations, further reducing labor costs while maintaining project schedules.

Operators can monitor the paving process in real-time and make adjustments remotely if necessary. This capability allows for immediate corrections if any discrepancies arise, ensuring the project stays on track.

While GPS-controlled paving machines represent a major advancement in automated paving technology, the next frontier in paving systems is the integration of robotics. Robotic systems are being developed to take over more complex and delicate aspects of the paving process, further increasing precision and efficiency.

I will give an overview of robotic paving systems. Robotic paving systems go beyond the automation of machine movement and incorporate robotic arms or autonomous vehicles to handle specific tasks in the paving process. These systems are particularly useful for applications where human workers would be exposed to high risks or where precision and speed are critical. One example of robotic integration in paving is the use of robotic arms to place and level materials, particularly in areas that are difficult for traditional machines to reach. These robotic systems are typically equipped with sensors and cameras to provide feedback on the quality of the surface, ensuring that the asphalt is laid evenly and without defects.

Another example involves autonomous vehicles that can transport materials from a central supply point to the paving machine. These vehicles can navigate the construction site without human intervention, ensuring a constant supply of materials to the paver while reducing the need for manual labor.

Several benefits of robotic paving systems are improved precision, reduced labour costs, minimised human error, increased safety, and 24/7 operation..

Robotic systems, especially those with advanced sensors and cameras, can detect minute discrepancies in the paving process and make adjustments with high precision. This capability results in smoother, more durable roads. By automating tasks such as material placement and transport, robotic systems can significantly reduce the number of workers required on-site. This leads to lower labor costs and fewer safety concerns. Robotics eliminates the potential for human error in the most delicate aspects of the paving process. For example, robotic arms can ensure that the asphalt is placed with exact consistency, which would be difficult for a human worker to achieve consistently.

Safety is a critical concern in construction, with workers often exposed to hazardous environments such as heavy machinery, moving vehicles, and high-traffic areas. Automated paving systems enhance safety by reducing human involvement in dangerous tasks and ensuring that machines are precisely controlled, even in complex or risky situations. Take, for example, the use of robotic systems in the paving of the I-35W bridge replacement project in Minneapolis. By using autonomous machines and robotic arms, workers were able to reduce their exposure to traffic and heavy equipment. Robots were tasked with handling specific aspects of the paving process, such as leveling and material placement, minimizing the need for workers to be on the ground near moving machines. This improved safety and also reduced the likelihood of accidents or injuries.

Robotic paving systems can perform tasks that would normally require workers to be in close proximity to moving machinery, thus reducing the risk of accidents and injuries. Robotic systems can operate continuously, even at night or in low-light conditions. This capability increases project throughput and allows for faster completion times.

Despite the numerous advantages, the implementation of automated paving systems also presents several challenges. One of the primary concerns is the cost of equipment. GPS-controlled paving machines and robotic systems are expensive to purchase and maintain, which can be a barrier for some contractors, especially in smaller projects.

Additionally, while automation can improve efficiency, it also requires highly skilled operators to set up and monitor the systems. These operators must understand the complexities of both the technology and the construction process to ensure that the systems operate correctly.

Another challenge is the need for continuous innovation and adaptation. The construction industry is constantly evolving, and as new materials and techniques are developed, paving systems must adapt to these changes. This requires ongoing investment in research and development, as well as collaboration between technology providers and construction companies.

The future of automated paving systems is incredibly promising. As technology continues to advance, it is likely that paving systems will become even more intelligent and efficient. For example, machine learning and artificial intelligence could enable paving machines to learn from previous projects and make predictive adjustments, further enhancing their performance.

Moreover, the integration of unmanned aerial vehicles (UAVs) or drones could provide additional support in the monitoring and surveying of construction sites. These drones could capture real-time data on the state of the road surface, helping to identify areas that need attention before the paving process begins.

In the long term, the evolution of automated paving systems may lead to fully autonomous construction sites, where robots and machines perform all tasks with minimal human oversight. This shift could improve the speed and quality of road construction and create a safer, more efficient work environment.

Robotic maintenance and repair have revolutionized crack sealing and pothole patching. As the global infrastructure deteriorates, particularly roads and highways, maintaining and repairing these critical transportation networks is becoming an increasingly urgent challenge. Traditional road maintenance techniques, though effective, often involve labor-intensive processes, long downtimes, and inherent safety risks for workers. The integration of robotics into road maintenance, such as crack sealing and pothole patching, has introduced a new era of repair technology. By automating these tasks, robots can complete repairs faster, more efficiently, and with greater safety for workers. This paper explores the use of robots for crack sealing and pothole patching, examining real-life examples and analyzing the potential benefits of robotic systems in road maintenance.

While these repairs are necessary, they often require manual labor, which can be slow and hazardous. Workers need to deal with dangerous road conditions, exposure to traffic, and heavy equipment, making road repair a high-risk job. The development of robotic systems for road maintenance aims to address these challenges, offering the potential for faster, safer, and more consistent repairs.

Robotic systems for crack sealing typically consist of automated machines equipped with sensors, cameras, and precision tools. These machines can detect cracks in the pavement, clean them, and seal them with a rubberized or bituminous material without requiring significant human intervention.

The process begins with the deployment of a robot that is equipped with various sensors to scan the road surface for cracks. Advanced imaging systems, such as high-definition cameras or infrared sensors, can be used to detect even minor cracks, ensuring that all potential problems are addressed. The robot can then autonomously navigate the road surface, following the cracks to their full extent.

Once the cracks are identified, the robot uses specialized tools to clean the cracks and prepare them for sealing. This cleaning process often involves the use of air jets or brushes to remove debris, dirt, and moisture from the crack. Afterward, the robot applies the sealing material, typically a rubberized asphalt or polymer-based product, using precision-controlled equipment. This automated sealing process is faster and more consistent than manual methods, reducing the risk of human error and improving the quality of the repairs.

One example of robotic technology for crack sealing is the ROADBOT, developed by researchers at the University of California, Berkeley. ROADBOT is a fully autonomous crack-sealing robot designed to improve the speed, precision, and safety of crack repair. It uses an array of sensors, including GPS, to navigate road surfaces and accurately identify cracks that need to be sealed.

The robot cleans and seals cracks using a system of high-pressure air and hot rubber, completing the task without the need for human workers to be close to moving traffic.

The ROADBOT has been tested in real-world conditions, showing its ability to autonomously travel along a road, identify cracks, and apply sealing materials with impressive precision. Its development marks a significant step toward reducing the time, labor, and risks involved in traditional crack-sealing operations. Moreover, because the robot can operate continuously without fatigue, it can complete the same tasks much faster than human crews, thus reducing downtime on roadways and ensuring faster repairs.

Potholes, which are depressions or holes in the roadway surface caused by weathering, traffic wear, or material degradation, are another common issue that plagues road infrastructure. Pothole repair is a complex and often expensive process, particularly on high-traffic roads or in urban areas where disruption must be minimized. Traditionally, pothole patching involves filling the hole with an asphalt mix, compacting it, and allowing the mixture to cure. However, this process can be time-consuming and typically requires traffic control measures to ensure the safety of workers.

Robotic pothole patching systems aim to streamline this process by automating the detection, filling, and compaction of potholes. These robotic machines can autonomously identify potholes, assess their size and depth, and apply the appropriate repair materials in real-time.

Robotic pothole patching systems are typically equipped with a range of sensors and cameras to identify potholes on the road surface. Once a pothole is detected, the robot automatically drives to the location, where it deploys a series of automated tools to repair the damage. These tools often include specialized nozzles for filling the hole with a mixture of asphalt or other materials, as well as compacting mechanisms to ensure that the patch is firmly set in place.

The robot is able to work autonomously, meaning it can perform pothole repairs at any time, including during off-peak hours when traffic is lighter. This reduces the disruption to traffic and speeds up the repair process, allowing more potholes to be addressed in a shorter duration of time.

A prominent example of a robotic pothole-patching system is PatchMaster, developed by the British engineering company Civil Systems Engineering. PatchMaster is a robotic pothole-patching vehicle that autonomously repairs potholes by filling them with a hot asphalt mixture. The machine uses cameras and sensors to detect potholes and automatically positions itself to repair them, eliminating the need for human workers to operate the machine.

PatchMaster is equipped with a robotic arm that applies the asphalt patch, followed by a compaction system that ensures the patch is tightly bonded to the surrounding surface. The machine operates autonomously, requiring minimal human intervention, which not only speeds up repairs but also improves worker safety by reducing their exposure to hazardous road conditions.

In pilot tests conducted in the United Kingdom, PatchMaster has demonstrated the ability to fix potholes in less than five minutes, significantly reducing the time required for repairs. This is a stark contrast to the traditional method, which can take several hours, particularly on busy roads.

The use of robotics for crack sealing and pothole patching offers several key advantages over traditional manual methods. Robots can perform crack sealing and pothole patching tasks much faster than human crews. This efficiency minimizes road closures and traffic disruptions, allowing for quicker turnaround times and less inconvenience for drivers. Robotic systems, equipped with advanced sensors and imaging technologies, can identify and address even the smallest cracks or potholes with high precision. This reduces the likelihood of overlooking smaller defects that could cause long-term damage to the road surface. Although the initial investment in robotic systems can be high, the long-term savings in labor costs, reduced downtime, and fewer repair failures can make robotic maintenance more cost-effective than traditional methods. Robots also operate efficiently, reducing the amount of repair material needed and minimizing waste. Road maintenance often involves working in hazardous conditions, such as on busy highways with heavy traffic. Robotic systems remove workers from dangerous situations by performing repairs autonomously. This reduces the risk of accidents and injuries on the job.

Despite the clear advantages, there are still several challenges to the widespread adoption of robotic systems for road maintenance. While robotic systems can save money in the long run, the initial cost of purchasing and maintaining the robots can be a significant barrier for some municipalities or contractors. Robotic systems still require constant refinement. Current robots may struggle with very complex or irregular surfaces, and they may need to be adapted for different types of pavement or weather conditions. The integration of robots into public infrastructure projects may face resistance from both workers and the general public. Some workers may be concerned about job displacement, while citizens might question the reliability or safety of robotic systems.

Robotic systems for crack sealing and pothole patching are transforming road maintenance, providing faster, safer, and more efficient solutions to common infrastructure problems. Real-life examples, such as ROADBOT and PatchMaster, illustrate the potential for robotics to improve the quality of road repairs, reduce costs, and enhance safety for both workers and drivers. As technology continues to evolve, robotic systems will likely become an increasingly important part of infrastructure maintenance, contributing to the creation of safer and more efficient transportation networks around the world.

Robots can work around the clock, reducing delays and increasing productivity. For example, potholes and cracks can be repaired at night when traffic is lighter, ensuring minimal disruption and faster completion times.

Thus, in the context of Construction 4.0, there is a potential to deliver personalized, intelligent, and interconnected building products. This necessitates a digital transformation within the industry, encompassing the digitalization of design, construction, and operations. Such a digital transformation entails a fundamental paradigm shift in the construction sector. This shift must occur across multiple dimensions, including technology, organization, and policy. Additionally, it necessitates a comprehensive integration of information, processes, knowledge, and personnel.

CHAPTER 8

BRIDGING THE GAP: TRADITIONAL WISDOM MEETS MODERN INNOVATION IN ROAD CONSTRUCTION

Todd Weyandt, an award-winning podcast host, writes in his article titled 'Innovation Champion for Construction', " Embracing more innovation in the construction industry requires a thoughtful approach that acknowledges the delicate balance between tradition and progress. Through strategic mentoring, an understanding of onsite realities and transparent communication, the industry can navigate the challenges and set new standards for success and sustainable growth. Those who embrace innovation with an open mind in the evolving construction sector will help shape the future of the industry."

When examining the evolution of traditions and practices, critical questions arise regarding which traditions to retain and which to leave behind, along with the rationale behind these decisions. In a rapidly changing world, societal values and technological advancements challenge the relevance of long-standing customs. It is essential to evaluate these traditions comprehensively, considering both their cultural significance and practical implications.

Traditional construction techniques have been honed over centuries, reflecting a deep understanding of local materials, environmental conditions, and community needs. These methods often emphasize using local resources and labor, which can lead to more sustainable practices. For example, in many regions, traditional earth-building techniques use local soil that can be both cost-effective and environmentally friendly. By combining these time-tested methods with modern technology, such as advanced materials, data analytics, and mechanized construction equipment, we can achieve optimal results that honor the past while embracing the future.

One key aspect of this evaluation is the need to shun redundant practices. As industries evolve, the availability of skilled labor often diminishes, leading to a shortage of expertise required to uphold certain traditional methods. Alongside this, the supply of materials may become inconsistent or prohibitively expensive. For instance, crafting practices that rely on specific, now-scarce resources may no longer be feasible, prompting a shift towards more sustainable and cost-effective alternatives. Innovations in asphalt use for road construction is the most suitable example.

Moreover, the pressing urgency of environmental challenges, including climate change and global warming, necessitates a transition towards sustainable practices. Traditional methods may not always align with contemporary environmental needs, and embracing new technologies becomes essential. This might entail adopting greener processes that reduce waste, conserve resources, and lower carbon footprints. For instance, integrating renewable energy sources and eco-friendly materials can enhance production while ensuring a commitment to ecological integrity. Using natural products like jute and recycled plastics in road construction are appropriate examples of innovation in traditional products.

Ultimately, the selection of which traditions to retain or abandon is a nuanced process. It involves balancing respect for cultural heritage with the practical realities of modern life and environmental sustainability. By thoughtfully considering the implications of our choices, we can forge paths that honour our history while embracing technologies and practices that are better suited to serve future generations. This reflection ensures that we preserve meaningful traditions while also equipping ourselves to tackle the challenges of the contemporary world.

Let me explain with an example from India. The Union Ministry of Road Transport and Highways has directed the Kerala State Government to incorporate rubber in road construction despite the higher costs, highlighting its benefits for passenger safety and the longevity of roads. Rubber is grown in large quantities there, and this would benefit the local people. The State Public Works Department estimates an additional cost of Rupees 297,418 for constructing a 1 km road with a rubber-mixed bitumen. The Ministry asserts that tests show roads with natural rubber significantly reduce accident risks and improve grip, ultimately decreasing maintenance costs and enhancing vehicle safety. Experts suggest that the long-term benefits justify the initial investment.

Let me now examine the real-life issues that are compelling the otherwise slow and traditional construction industry to look for innovative alternatives and support systems. Some common impediments that can be counted on the fingertips are a shortage of skilled labour, a lack of resources, overdue projects, slow delivery, and high costs. Also, the geographical demands are varied. For example, the requirements of building roads in Indian villages are diagonally opposite to those of constructing roads in cities in Qatar.

I will address these issues by explaining how automation and digitization have led to innovations and solved these problems to a great extent. I will also dwell upon the challenges.

A major factor to consider in this evaluation is the need to eliminate outdated and redundant practices. The construction industry is grappling with a diminishing pool of skilled labor, which directly impacts its ability to continue certain traditional methodologies. Additionally, fluctuations in material availability may render some practices impractical or costly. For example, construction techniques that depend on rare materials may no longer be viable, pushing the industry to seek more sustainable and economically feasible alternatives. A prime illustration of this shift can be observed in the innovative approaches to asphalt utilization in road construction.

Environmental challenges, such as climate change and global warming, further underscore the need for more sustainable practices. Conventional construction methods may not address modern ecological demands adequately, highlighting the necessity of incorporating new technologies. This transition may involve adopting environmentally friendly processes that minimize waste, conserve resources, and lower carbon emissions. The use of natural materials such as jute and recycled plastics in road construction exemplifies how traditional methods can evolve to meet contemporary sustainability goals.

Ultimately, deciding which traditions to preserve or discard is a complex issue that requires a nuanced understanding. It involves balancing respect for cultural heritage with the practical needs of modern life and ecological sustainability. By thoughtfully considering the consequences of our choices, we can create pathways that honor our past while embracing advancements that are better suited for the challenges ahead. This reflective approach ensures that we maintain valuable traditions while also preparing to address the demands of the modern world.

Summarizing why the construction industry, traditionally slow to change, is now seeking innovative alternatives, we must examine the real-life issues at play. Among the most pressing challenges are:

- A shortage of skilled labor

- Insufficient resources

- Delayed projects

- Slow delivery times

- Escalating costs

Moreover, the geographical context significantly influences construction needs. For instance, the requirements for road construction in rural Indian villages starkly contrast with those in urban centers like Qatar.

In addressing the aforementioned issues, it's essential to explore how automation and digital innovations have begun to transform and alleviate these challenges. Automation and AI are revolutionizing various aspects of the construction process, helping to mitigate labor shortages while maintaining high-quality standards.

The labor shortage within the construction sector is a pressing issue caused by a variety of factors. Key contributors include an aging workforce, with many skilled workers nearing retirement, and a lack of enthusiasm among younger generations to pursue careers in construction. Compounding this, the demand for infrastructure, driven by rapid urbanization, continues to rise, putting further strain on an already limited labor pool.

To tackle the labor shortage, automation and AI-driven technologies are emerging as effective solutions. These advancements are changing the face of the construction industry by enhancing efficiency and safety while maintaining quality standards. One of the critical areas where AI is making a significant impact is project management. AI-driven systems can handle various aspects of project oversight, from scheduling to resource allocation, allowing firms to adapt quickly to changing demands.

Construction companies are leveraging automated machinery for tasks traditionally performed by human labor. This technology not only addresses the shortfall of skilled workers but also accelerates construction timelines and reduces costs. As a result, businesses can maintain competitive pricing while enhancing safety through the reduction of human error.

Moreover, integrating AI into training and skill development programs can bridge the gap left by retiring workers. AI can facilitate personalized training experiences, allowing new workers to gain the skills needed efficiently. Through virtual reality and simulation technologies, trainees can practice real-world scenarios safely, preparing them for the challenges they will face on the job site.

The rising demand for commercial real estate and infrastructure projects means that embracing AI and automation will be crucial for sustained growth in the construction sector. Companies that adopt these technologies early will not only mitigate current labor shortages but will also position themselves favorably in a rapidly transforming industry.

In short, I can say, the construction industry stands at a crossroads where it must find a way to combine venerable traditions with modern innovations. By critically evaluating which practices to retain and to what extent or replace, and by integrating automation and AI, the industry can effectively address labor shortages and adapt to the pressing demands of today's environment. Embracing these changes entails acknowledging historical methods and recognizing the need for evolution, blending the best of both worlds to ensure a sustainable future in construction.

So far, we have understood that bridging the gap between traditional road construction methods and modern innovations is crucial for creating more sustainable, efficient, and resilient infrastructure. This involves integrating smart technologies, like IoT sensors and AI, with established practices to enhance communication, safety, and construction efficiency. At this juncture, it is important to discuss innovations in water-saving pavements and 3D printing that are playing a key role in reshaping the future of road construction.

The rise of smart roads has brought a revolution in the field of road construction. As I have discussed in great detail in the previous chapter, roads are becoming increasingly equipped with sensors, IoT devices, and AI to monitor traffic patterns, road conditions, and even generate electricity through solar roads. Real-time data allows for proactive maintenance, traffic management, and potentially even self-healing pavements. For example, solar roads that generate power, smart sensors that monitor road conditions, and AI-powered traffic management systems are all examples of this trend.

With the coming up of smart roads, embracing sustainable practices has become the need of the hour. Water-saving pavements reduce stormwater runoff and increase road durability, contributing to environmental sustainability. Cool Pavements are modified pavements that reflect solar energy and enhance water evaporation to stay cooler than traditional pavements. Using recycled materials, implementing energy-efficient construction practices, and minimizing waste are also crucial aspects of sustainable road construction.

Leveraging technological advancements is essential. 3D printing technology is revolutionizing the construction industry, including road construction, with applications like 3D concrete printing. Drones are used for site surveying, monitoring construction progress, and inspecting infrastructure. BIM helps to visualize and manage the entire project lifecycle, from design to maintenance.

The importance of collaboration is increasing with increasing need to curb wastage and conserve time. Modern technology plays a crucial role in the integration process. Innovations such as 3D printing, drone surveying, and smart materials allow for greater precision and efficiency in construction. Moreover, data-driven decision-making enables engineers to analyze patterns and optimize designs, significantly reducing waste and time. For instance, Geographic Information Systems (GIS) can help identify the most effective routes for new roads by analyzing terrain and environmental impact before construction begins.

Collaboration is key in harnessing the best aspects of both traditional and modern approaches. The successful implementation of innovative road construction practices hinges on the cooperation between engineers, scientists, and skilled construction workers. Each group brings unique insights and competencies to the table. Engineers provide the technical expertise required for modern construction practices, while scientists may offer insights into material properties and environmental impacts. On the other end, construction workers often possess valuable knowledge about local practices and the capabilities of different materials. Workers may have practical knowledge about which types of local aggregates (sand, gravel, crushed stone) are most suitable for the specific soil conditions at a construction site. This knowledge can be invaluable in optimizing road construction, preventing issues like cracking or erosion, and ensuring the road's longevity.

Workers often develop a deep understanding of how different materials perform in various situations, including the effects of weather, traffic, and seasonal changes. For instance, they might know which asphalt mixes are best suited for handling the hot summers and occasional heavy rains of South East Delhi or that of a city in South India with heavy and perennial rainfall. This synergy can lead to more efficient construction processes and higher-quality outcomes.

Moreover, involving local communities in the planning and construction phases is essential for creating road infrastructure that truly meets the needs of society. By leveraging traditional knowledge, especially from indigenous communities or those with a rich history of local construction practices, we can ensure that new developments are well adapted to their environments and culturally sensitive. This collaborative approach not only builds trust but also fosters a sense of ownership among community members, enhancing the longevity of constructed infrastructures.

As we look to the future, the road construction industry must continue to embrace emerging trends that emphasize resilience and sustainability. Climate change poses unprecedented challenges, urging us to create infrastructure that can withstand extreme weather events and adapt to shifting environmental conditions. By integrating traditional methods and modern technologies, we can build roads that are not only functional but also environmentally friendly and capable of supporting community needs for generations to come. I will explain this point by writing in detail about the cool asphalt technologies.

In conclusion, the future of road construction lies in the harmonious blend of traditional knowledge and modern technology. By fostering collaboration among various stakeholders and emphasizing sustainable practices, we can build a more resilient and efficient infrastructure. This integrated approach promises not just better roads, but a more sustainable future for our communities. As the industry evolves, it will be crucial to keep these principles at the forefront of our efforts.

The synergy between old and new approaches is the key to success. Road construction exemplifies the remarkable synergy achieved by merging time-tested practices with cutting-edge innovations. For instance, by incorporating recycled materials like waste plastic, we can significantly enhance road performance while promoting sustainability. Furthermore, implementing robust road safety measures from the design phase ensures that our infrastructure is both safe and effective, while technologies such as Building Information Modeling (BIM) streamline project management for improved efficiency.

Consider India's groundbreaking "1000 km Plastic Tar Road" initiative, which successfully integrated waste plastic into the bituminous mix. This initiative not only showcases the potential of recycled materials in construction but also sets a precedent for sustainable practices in the industry.

In Karnataka, the "Safe Roads for Development (SR4D)" initiative stands as a testament to the power of safety engineering. By conducting thorough safety audits and incorporating strategic road-safety measures, the project fostered a safer road network without straining the budget—a win-win for communities.

Moreover, IndiaRAP's case studies reveal the transformative impacts of improved intersections. Enhancements such as additional right-turn lanes, refuge areas, and elevated pedestrian crossings significantly bolster safety and efficiency for all users of the road.

The modern technique of Full-Depth Reclamation revolutionizes road repair by blending old asphalt with underlying gravel. This innovative approach minimizes waste while maximizing the use of existing materials.

To further enhance road stability, geotextiles—particularly woven types—play a crucial role in stabilizing the road base. They prevent the mixing of aggregate layers, ensuring a solid foundation that stands the test of time. This combination of traditional methods with modern materials and techniques underscores a commitment to quality and durability.

Traffic management is entering a new era with systems like the Coordinated Adaptive Traffic System (SCATS) in Sydney. This technology dramatically improves traffic flow and safety, paving the way for similar implementations worldwide.

Lastly, Building Information Modeling (BIM) revolutionizes project management by acting as a centralized hub for design, construction, and maintenance data. This promotes collaboration, enhances transparency, and drives efficiency throughout the entire project lifecycle.

In summary, by harnessing these innovative strategies and technologies, we can redefine the future of road construction, creating safer, more efficient, and sustainable infrastructure that meets the needs of our communities today and in the future.

One of the major fields where innovation in road construction is required is in UHIs. Urban Heat Islands (UHIs) are metropolitan areas that experience significantly higher temperatures than their rural surroundings, primarily due to human activities and alterations in land surfaces. One of the major contributors to UHI effects is traditional asphalt pavement, which absorbs and retains heat. To combat this issue, cool asphalt technologies have emerged as an effective mitigation strategy.

Cool asphalt technologies involve the use of materials and methods designed to decrease surface temperatures and reflect more sunlight. These techniques can include the incorporation of lighter-colored aggregates, reflective coatings, or special polymers in asphalt mixtures. By enhancing the albedo, the surface's ability to reflect sunlight, these materials can reduce the amount of heat absorbed during the day, leading to cooler surface temperatures and diminishing the overall heat contribution to the surrounding environment.

The benefits of implementing cool asphalt solutions extend beyond mere temperature reduction. By mitigating UHIs, these technologies can improve energy efficiency in nearby buildings, as cooler outdoor conditions can lessen the reliance on air conditioning systems. This can lead to lower energy consumption and reduced greenhouse gas emissions. Furthermore, cooler asphalt can enhance urban comfort for pedestrians and cyclists, encouraging outdoor activities and promoting healthier lifestyles.

Additionally, cool asphalt can play a role in prolonging the lifespan of pavements. Traditional asphalt can suffer from thermal cracking due to extreme temperatures, while cooler alternatives may experience less stress from heat, potentially reducing maintenance costs. The adoption of cool asphalt technologies represents a sustainable approach to urban planning and infrastructure development, contributing to the reduction of UHI effects. By embracing these innovations, cities can create a more comfortable and environmentally friendly living environment for their residents.

It is important to explore the potential benefits and costs of cool pavements. Cool pavement technologies have gained significant attention as urban areas continue to grapple with the challenges of climate change, urban heat islands (UHI), and the increasing demand for sustainable infrastructure. Implementing cool pavements—materials designed to reflect rather than absorb solar energy—can offer a range of benefits, while also presenting some potential costs.

Urban areas tend to be significantly warmer than their rural surroundings, primarily due to human activities and the prevalence of heat-absorbing materials like asphalt. Cool pavements can help lower surface and ambient temperatures, thus mitigating the UHI effect. By reducing the heat retained in urban areas, cool pavements can contribute to more comfortable living conditions during hot months.

As temperatures drop due to cooler pavements, there is a diminished need for air conditioning in buildings, leading to reduced energy consumption. This reduction translates into lower greenhouse gas emissions as less fossil fuel-based electricity is generated, aligning with global efforts to combat climate change.

Certain types of cool pavements, particularly permeable pavements, allow water to infiltrate through the surface, effectively reducing runoff. This capability not only helps in managing stormwater but also minimizes the risk of flooding. Additionally, as rainwater passes through these pavements, it can be naturally filtered, improving water quality.

Cool pavements are often designed to be more durable than traditional options. Their reflective properties can reduce thermal cracking, allowing pavements to last longer and thus generating less waste over time from repairs or replacements.

Longer-lasting pavements naturally lead to decreased maintenance requirements. By reducing the frequency of repair and replacement work, municipalities can allocate funds to other important infrastructure projects, enhancing overall urban sustainability.

Cool pavements can reflect more light, enhancing visibility at night without the need for additional lighting infrastructure. This property can improve safety for pedestrians and vehicles alike, leading to fewer accidents.

Cooler temperatures not only enhance the comfort of pedestrians and residents but can also improve outdoor recreational spaces. Parks and public areas with cool pavements will be more attractive for social gatherings and community activities.

Permeable cool pavements that effectively manage water runoff can significantly increase driving safety by reducing hydroplaning risks and improving traction during wet conditions.

Some cool pavement materials may contribute to noise reduction, making urban areas quieter and more pleasant environments to live in.

While the benefits of cool pavements are compelling, there are also potential costs associated with their implementation. Initially, the installation of cool pavement technologies can be more expensive than traditional asphalt or concrete due to the materials and specialized processes involved.

Moreover, there may be concerns about the long-term performance of cool pavements, particularly in terms of their effectiveness in cooler climates or areas with heavy traffic loads. Some cool pavements may also require more extensive maintenance than traditional options, which could offset some of the financial benefits related to reduced maintenance costs.

Additionally, the initial implementation may necessitate a change in design practices and regulations, leading to further investment in training for professionals involved in construction and urban planning.

Thus, while cool pavements present a progressive solution to urban heating and environmental challenges, careful consideration must be given to both their benefits and costs. Thoughtful implementation can lead to healthier urban environments and contribute positively toward climate resilience.

A growing number of cities around the globe are taking bold steps to combat the urban heat island effect and address climate change through innovative cool pavement solutions. Notable examples include Doha, Qatar, Los Angeles, Phoenix, and Rome, with Athens, Greece, leading the way in the extensive use of cool pavement materials for urban rehabilitation.

In Doha, a groundbreaking pilot project is underway, utilizing a cryogenic material designed to significantly lower asphalt temperatures. Meanwhile, Los Angeles is actively experimenting with cool pavements, having launched an impressive initiative to coat 200 blocks of neighborhood streets with reflective sealants. Phoenix, too, is making strides with a large pilot project, incorporating cool paving as a vital strategy to tackle climate change across more than 100 miles of paved streets. Other cities, including San Antonio, Nashville, and Davis, are also exploring and implementing their cool pavement projects, according to the American Society of Civil Engineers (ASCE).

Athens stands out, having employed cool colored paving materials in a sweeping urban rehabilitation initiative, demonstrating the effectiveness of these solutions in mitigating heat. Additionally, pilot studies in cities like Tirana (Albania) and Sharjah (UAE) are further exploring the potential of cool pavements, as highlighted by ScienceDirect.com. Rome has witnessed remarkable reductions in radiant temperatures thanks to high-reflective pavements that provide much-needed relief from the heat.

These initiatives are not just a trend; they represent a crucial shift towards sustainable urban development that can improve livability and combat the harsh realities of climate change. It's time for more cities to embrace these innovative solutions for a cooler, healthier future.

Cool pavements are a powerful solution to combat rising urban temperatures, and they come in two innovative forms: solar-reflective coatings and permeable asphalt. When strategically implemented, these technologies can significantly alleviate heat in our cities through unique mechanisms.

Solar-reflective coatings, typically lighter in color, boost the pavement's albedo. By reflecting more solar radiation, they minimize heat absorption, effectively keeping surfaces cooler. These coatings utilize advanced materials such as reflective pigments and high-performance binders that maximize solar reflectance. As a result, they not only lower surface temperatures but also contribute to reduced energy consumption for cooling buildings and enhance nighttime visibility.

On the other hand, permeable asphalt operates through the natural process of water evaporation, utilizing the latent heat of vaporization to further cool the environment. This dual approach—reflecting sunlight and enhancing moisture evaporation—creates a comprehensive strategy to mitigate urban heat.

While the benefits are substantial, it's important to acknowledge that the implementation of reflective coatings may pose challenges regarding durability and skid resistance, along with potential aesthetic considerations. Despite these hurdles, the advantages they offer in creating more livable, cooler urban spaces cannot be overlooked. Investing in cool pavement solutions is not just a trend;

it's an essential step toward sustainable urban development and a healthier environment for all.

Permeable asphalt is a revolutionary solution that allows water to seep into the pavement, where it can evaporate and effectively cool the surface. This innovative process not only lowers temperatures but also promotes a more sustainable urban environment. Engineered with porous aggregates and specialized binders, permeable pavements create a natural reservoir for water, enhancing evaporation and combating the heat generated in urban areas.

The benefits extend beyond just temperature reduction; permeable pavements play a vital role in improving water management by minimizing runoff and significantly lowering the risk of flooding. The availability of moisture is essential for this evaporative cooling to work effectively, making these pavements particularly beneficial in areas with sufficient rainfall or optimal evaporation conditions.

In essence, both reflective coatings and permeable asphalt are powerful assets in the fight against rising urban temperatures. By strategically implementing these technologies, cities can tackle the urban heat island effect head-on, fostering more comfortable, resilient, and sustainable environments for all. Embracing these advancements is not just an option; it's a necessity for the future of urban living.

CHAPTER 9

GREEN FOUNDATIONS: BUILDING A SUSTAINABLE FUTURE

When I mention green foundations in the context of road construction, I mean using sustainable materials and practices to minimize the environmental impact of road building. This includes utilizing recycled asphalt pavement (RAP), recycled asphalt shingles (RAS), warm-mix asphalt (WMA), and bio-based binders, thereby reducing carbon emissions and conserving natural resources. Green road construction incorporates the entire road lifecycle, from design and construction to maintenance and disposal, aiming for a more sustainable and efficient infrastructure.

In the previous chapters, I have discussed the technical innovations in road construction, the Industrial Revolution 4.0, automation, digital systems, and designs. In this chapter, I will describe innovations in asphalt mixtures used as the main ingredient in road construction.

An important question must be answered: why are these innovations important for sustainability and the environment?

Beyond the obvious, here are some insights to explore sustainable materials. The advantages of eco-friendly road construction reach well beyond what meets the eye. From better water runoff management to increased safety, every aspect contributes to developing more sustainable and resilient infrastructure.

Innovative material research is the key to responsible paving. The exploration of new materials plays a critical role in eco-conscious road construction. Utilizing options such as permeable pavements and recycled aggregates transforms how we create roads, ensuring durability while minimizing environmental impact.

There is a global commitment to lowering carbon emissions. One of the main objectives of green road construction is to diminish the carbon emissions associated with conventional practices. Sustainable bitumen, known for its reduced carbon output, is pivotal in fulfilling this goal.

Economically viable and environmentally friendly approaches are the need of the hour. It's a common misconception that sustainable methods are expensive. In reality, green road construction demonstrates that environmentally friendly practices can also be budget-friendly. Investing in eco-sustainability doesn't have to mean exceeding financial limits. Recycling on-site debris is a budget-friendly practice.

In the majority of the construction projects, especially highway construction projects, natural rock material such as crushed rock, broken stone, sand and gravel is used. Recycled materials can partially or even fully replace these natural materials. Recycling materials like asphalt and concrete reduces the need for virgin materials, lowers emissions, and can be cost-effective, as the materials are already on-site. Similarly, incorporating recycled plastic, rubber, or waste materials like tyres or fly ash in pavement and other construction elements can reduce material costs and environmental impact.

France, for example, built the world's first fully recycled motorway using 100% recycled asphalt. Spain has also used recycled rubber crumbs from tires in over 1,600 kilometers of roads. The Basque Country in Spain even used 80% recovered material from aging roads for a highway. Additionally, many countries use recycled materials like crushed concrete, fly ash, and shredded plastics in asphalt mixes.

Engaging communities in green road initiatives has proved to be helpful in both developing and developed countries. The development of green roads transcends mere construction; it's a collaborative effort with local communities. Including residents in the decision-making process creates a sense of ownership and responsibility, making these roads vital to their lives.

The "Green Highway" project in northern Sweden and Norway is a notable example of a collaborative effort. It's a 450-km-long, fossil-free transportation corridor that utilizes renewable energy and promotes eco-friendly transportation along the E6 and E14 motorways. The project emphasizes reducing CO2 emissions, connecting low-density towns, and supporting local economies through renewable energy.

Like any other initiative, navigating challenges in sustainable road construction is inevitable. Though the transition to eco-friendly road construction is encouraging, it is challenging. Overcoming technological barriers and raising public awareness are essential for the broad acceptance of green initiatives. The stakeholders need to understand that immediate high costs may be alarming, but in the long term, they are more beneficial and compensate for the initial high costs.

Various global efforts have successfully adopted green road construction practices. Analyzing these success stories provides valuable lessons on effective strategies and how we can adapt them to different environments. Examples include the use of recycled asphalt in pavements, the implementation of solar roadways, and the incorporation of green infrastructure like wildlife corridors. The United States is involved in establishing green corridors with South Korea and is also a partner in the Global South Green Corridors project. Singapore and Australia are collaborating on the Singapore-Australia Green Digital Corridor. Japan is also involved in green shipping corridors. Brazil, Paraguay, Argentina, and Chile are involved in a green corridor connecting these countries.

As technology and sustainable practices progress, the future of eco-friendly road construction looks bright. Innovations such as smart roads and advanced materials offer an array of possibilities to redefine our transportation infrastructure.

In summary, moving toward sustainable road construction is not merely an option; it's a duty we owe to the planet. By adopting eco-friendly practices, particularly through sustainable bitumen, we can ensure that our roads endure and contribute to a healthier and greener Earth.

NAPA is shifting from a traditional linear model of "take-make-dispose" to a more resource-efficient and environmentally friendly circular economy in road construction. Key initiatives include advocating for the use of Recycled Asphalt Pavement (RAP) to minimize the need for virgin materials and reduce carbon emissions and energy consumption. Their "*The Road Forward*" initiative aims for net-zero greenhouse gas emissions throughout the asphalt pavement life cycle while promoting sustainability and circularity. NAPA emphasizes resource efficiency, encouraging construction practices that lower raw material and energy use, as well as designing pavements for future disassembly and reuse to extend their lifespan. Furthermore, NAPA promotes collaboration among industry stakeholders and actively shapes policies and regulations that support circular economy principles in transportation.

Having talked about the need and scope for green roads, let me explore what needs to be done.

The asphalt industry faces increasing pressure to reduce its environmental impact and improve economic efficiency. Warm mix asphalt (WMA) technologies have been developed as a promising solution to reduce the production temperatures of asphalt mixtures, thus improving energy efficiency and minimizing emissions. Besides, incorporating reclaimed asphalt pavement (RAP) material offers significant environmental and economic benefits by reducing the demand for virgin materials. However, high RAP contents present challenges such as binder hardening and poor adhesion among aggregates due to aging, limiting their application. This review delves into the interplay between WMA technologies and RAP, exploring their individual and combined effects on the environmental and mechanical performance of asphalt mixtures. By optimizing the selection and dosage of WMA additives and balancing RAP content, it is possible to mitigate the challenges associated with high RAP content and achieve the desired pavement performance. Future research should focus on developing advanced WMA additives and techniques to further enhance the integration of high RAP content, ensuring environmental sustainability and improved pavement performance.

Reclaimed asphalt pavement (RAP) is essential for promoting sustainability in highway engineering. Standard pavement repair methods, like crack sealing, patching, and overlaying, typically involve milling the pavement, which produces large amounts of recyclable material suitable for RAP mixtures. RAP is made from asphalt mixtures derived from milling existing road surfaces and from surplus or substandard products. This definition sets RAP apart from asphalt mixtures removed from different bituminous layers, as seen in countries such as the United States and Australia. RAP is designed to be incorporated immediately into new asphalt mixtures and requires further processing steps like crushing, granulation, and homogenization.

Recycled Asphalt Shingles (RAS) present an innovative and eco-friendly solution for reusing roofing materials by incorporating them into new asphalt mixtures. This method helps in minimizing waste by diverting materials from landfills and promotes resource conservation, making it a practice that is both environmentally responsible and economically advantageous. The adoption of RAS is increasingly recognized for its numerous benefits, including significant sustainability improvements. By recycling RAS, we can effectively utilize a material that would otherwise contribute to landfill waste, significantly reducing the environmental footprint associated with roofing materials. Furthermore, integrating RAS into asphalt production conserves valuable resources by decreasing the reliance on virgin asphalt and aggregate, which are essential components in paving. This practice can also lead to financial savings, as RAS is often less costly than traditional raw materials, decreasing overall production costs in the asphalt industry.

Beyond economic benefits, RAS can enhance pavement performance. The unique properties of RAS, such as its angular fine aggregates and fiber content, can improve the overall quality and durability of asphalt mixtures while potentially increasing their stiffness. This increased stiffness, however, does come with its own set of considerations. The binder derived from RAS tends to be stiffer than standard virgin asphalt binder, which can influence how the final asphalt mixture performs under various conditions. Additionally, the angular shape of RAS granules can elevate the voids in mineral aggregate (VMA), potentially affecting the final product's density and durability. To harness the full potential of RAS, careful blending with virgin materials and thoughtful design adjustments are essential. This ensures the asphalt mixture achieves its intended application's desired performance and longevity.

It's vital to recognize that while RAS can offer financial advantages, the costs associated with the recycling process and the incorporation of RAS into asphalt mixtures must also be factored into the overall budget. Two primary types of RAS can be recycled for use in asphalt: Tear-Off Asphalt Shingles (TOAS) and Manufactured Waste Asphalt Shingles (MWAS).

TOAS are typically gathered from residential and commercial buildings during roof repairs or replacements, whereas MWAS is generated during the manufacturing process of asphalt shingles. These materials represent a readily available and useful resource that can effectively contribute to more sustainable construction practices. As the construction industry continues to focus on sustainability and reducing environmental impacts, the use of RAS in asphalt production is poised to become an increasingly popular practice. By engineering RAS mixtures thoughtfully, it is possible to customize the characteristics of the asphalt to meet the specific needs of various projects, enhancing flexibility and adaptability.

In conclusion, the use of Recycled Asphalt Shingles in asphalt production not only offers a sustainable alternative but also promotes economic benefits, improved pavement performance, and design flexibility. However, careful consideration of factors such as stiffness, void content, blending techniques, and associated recycling costs is essential to ensure the successful implementation of RAS in paving materials.

The integration of RAS represents a significant step towards a more sustainable future in construction, allowing us to set a precedent for responsible resource use and waste reduction. By embracing and advancing practices such as the recycling of asphalt shingles, the industry can contribute meaningfully to environmental preservation while also meeting the growing demand for high-quality paving solutions.

The US Department of Transportation (USDOT), through its Federal Highway Administration (FHWA), promotes the use of Reclaimed Asphalt Pavement (RAP) and Reclaimed Asphalt Shingles (RAS) in asphalt mixtures due to their cost savings, environmental benefits, and resource conservation. By replacing virgin materials, the use of RAP and RAS significantly lowers asphalt mixture costs, as evidenced by substantial savings in 2021 alone. Environmentally, these practices help conserve natural resources and reduce landfill waste, aligning with USDOT's focus on environmental stewardship. The FHWA supports this initiative by conducting research, developing guidelines, and sharing best practices to enhance RAP/RAS mixture performance, including addressing challenges like low-temperature cracking. Many state agencies have successfully incorporated high percentages of recycled materials, with Florida utilizing RAP in over 75% of its asphalt mixtures, while ongoing research continues to optimize the design and performance of these sustainable materials.

A Life Cycle Assessment (LCA) for asphalt pavements is a comprehensive evaluation designed to analyze the environmental impacts at every stage, from the extraction of raw materials to the disposal of asphalt. This process aims to provide a holistic understanding of the environmental footprint associated with asphalt pavements. It involves a careful scrutiny of various factors, including resource utilization, energy consumption, and emissions released into the atmosphere throughout several key life cycle phases: material extraction, construction, maintenance, usage, and eventual disposal. By adopting this framework, stakeholders can gain insights into the overall sustainability of asphalt pavements and identify areas for improvement to mitigate adverse environmental effects.

There has been a significant rise in interest in comparing Life Cycle Assessments (LCAs), potentially linked to upcoming requirements for road infrastructure that will demand more resources in construction processes by 2050, leading to increased emissions. Consequently, scientific research focused on minimizing impacts has seen ongoing advancements. Additionally, the emphasis and guidelines from road authorities regarding documentation and the objectives for reducing greenhouse gas emissions further amplify these effects.

The LCA process is systematically divided into specific stages that guide the analysis. The first step is goal and scope definition, wherein the purpose and boundaries of the assessment are articulated. This includes identifying the asphalt pavement under study, establishing a functional unit—for example, 1 square meter of pavement—and specifying the geographical context. Following this, the Life Cycle Inventory (LCI) stage involves gathering comprehensive data about all inputs and outputs across the life cycle phases, covering aspects such as materials, energy usage, and associated emissions.

Next, the Life Cycle Impact Assessment (LCIA) translates this inventory data into a characterization of relevant environmental impacts through standardized categories, enabling a clearer understanding of issues like climate change due to greenhouse gas emissions from the various lifecycle stages. The final phase, Life Cycle Interpretation, focuses on analyzing the results obtained from the LCIA, identifying key environmental hotspots, and assessing the overall ecological performance of the asphalt pavement. This structured approach ensures that all relevant factors—including climate change, resource depletion, human and eco-toxicity, water use, and waste generation—are considered thoroughly, allowing for informed decision-making regarding asphalt pavement design and management.

Addressing environmental impact categories during the LCA is vital for promoting sustainable practices in the asphalt industry. Climate change implications arise from greenhouse gas emissions linked to material extraction, production, and usage. Resource depletion concerns focus on the consumption of finite natural materials, such as aggregates and bitumen. Furthermore, human and eco-toxicity implications relate to the potential harm caused by emissions and leachate, necessitating careful monitoring and management. Water use is another critical aspect, with water consumption occurring during material production, construction processes, and ongoing maintenance activities. Lastly, waste generation must be evaluated, specifically the impacts of disposal or recycling options available at the end of the asphalt pavement's life cycle.

When conducting an LCA, several factors deserve particular attention to enhance sustainability outcomes. One important aspect is recycling, which offers significant environmental benefits by reusing asphalt pavement materials as opposed to relying solely on virgin inputs. Additionally, the exploration of sustainable materials—like recycled plastics or other waste-based options—can contribute positively to the overall life cycle footprint, supporting a more circular economy approach. Maintenance practices also warrant evaluation, as different strategies can lead to varying environmental impacts over time. Finally, the assessment of end-of-life options is crucial, where the environmental consequences of multiple disposal methods—such as landfilling, incineration, or recycling—need to be understood thoroughly.

By executing a thorough Life Cycle Assessment, it becomes possible to pinpoint specific areas where strategies concerning asphalt pavement design and management can be enhanced to reduce environmental impacts and promote greater sustainability within the construction sector. This multifaceted perspective not only fosters innovation in the materials used but also encourages the adoption of practices that align with global sustainability goals, thereby contributing to a healthier planet.

Life Cycle Assessment (LCA) has been utilized to compare products or processes to aid decision-making aimed at reducing emissions without merely shifting the problem to other stages in the product's life cycle. However, for road projects, achieving this has proven challenging without undertaking a thorough LCA. This process demands significant time and resources, and the industry would greatly benefit from the ability to compare past and future evaluations to facilitate overall emission reductions during the decision-making process. Although LCA is regarded as a standardized approach, the findings indicate that this is not true for road projects.

A review of road design studies reveals that 82% lack transparency, complicating comparisons. Key parameters are often poorly specified, limiting reproducibility and effective emission reduction solutions. Most studies focus on new materials, with little attention to operation and maintenance emissions.

To promote sustainability in road infrastructure, it's vital to enhance the transparency and reproducibility of outcomes. By comparing different technological and material choices, road owners can better manage emissions over the road's lifespan through accurate information. We propose achieving this transparency by specifying key parameters, including detailed descriptions of road components related to the functional unit (FU), inventory databases, and impact assessment methods.

The study's scope should outline the dimensions of the road and its components, potentially through cross-sectional diagrams indicating the width and thickness of layers. Additional figures could detail elements like barriers or lighting systems, especially if the road features varying cross-sections. Each component must have its FU defined, linking it to the overall road's FU. Clear inventory data should connect each component's material flow to global warming potential scores. Lastly, the impact assessment methods must cover environmental indicators such as GWP and CED for components and the road, enabling comparisons across different projects.

This component-based analysis enhances usability and transparency, aiding LCA practitioners and the broader road construction sector in selecting materials and processes to achieve a 50% reduction in GHG emissions by 2050.

Perusal of many systematic literature reviews highlighted key knowledge gaps requiring further investigation. Specifically, road design parameters are insufficiently detailed aspects affecting the thickness of road courses and the materials used, impacting environmental outcomes for both flexible and rigid pavements. There is a scarcity of studies on rigid pavement rehabilitation, signaling a need for exploration. Concerns about result reliability, particularly regarding uncertainty analysis often neglected in research, also arise. Lastly, comprehensive uncertainty analyses on input data have yet to be conducted.

In conclusion, embracing green foundations is essential for creating a sustainable future. By prioritizing eco-friendly practices and innovative solutions, we can mitigate environmental challenges and foster a healthier planet. Collective action and commitment to sustainability will pave the way for resilient communities and flourishing ecosystems. Let us invest in green initiatives today for a brighter tomorrow.

CHAPTER 10

THE ARCHITECT OF CHANGE: LEADERSHIP AND INNOVATION IN CONSTRUCTION

Innovation in road construction is crucial for sustainable development. Embracing new technologies and materials can enhance road durability, reduce environmental impact, and improve safety. Smart infrastructure solutions, such as intelligent traffic management and eco-friendly paving materials, can streamline operations and promote efficient transportation networks, addressing the growing challenges of urban mobility.

As part of my applied research and innovation in infrastructure technology, I designed and developed a demo system called **PaveIQ** — a smart decision-support tool for pavement condition monitoring and predictive maintenance.

PaveIQ integrates field-collected road condition data (such as Pavement Condition Index scores) with machine learning models to forecast future degradation patterns. What makes the tool unique is its ability to **visualize condition forecasts over geospatial maps**, allowing road agencies and paving contractors to **prioritize repairs proactively** rather than reactively.

It features a dynamic dashboard that supports Predictive analytics using historical road performance data, Real-time visual heatmaps of high-risk segments, and Maintenance planning with time-based and cost-efficiency indicators.

By empowering users to optimize road interventions **before failures occur**, PaveIQ aims to reduce lifecycle costs, minimize disruption, and enhance overall infrastructure resilience. It bridges a critical gap between AI research and real-world asphalt management.

This innovation builds on my broader work in integrating AI and automation into civil infrastructure — a central theme throughout my career.

Two more tools (mobile applications in both android & ios) I have mentioned here.They are fully done and have been used for over 5 years. The BrewPoint and Auto-shopkeeper were developed at Superior Paving Corp. While employed (still am) at Superior Paving Corp., I solely conceptualized, designed, developed, and maintained a field operations mobile application called BrewPoint. The app was created to streamline plant-to-jobsite coordination and improve real-time communication among field crews and supervisors. BrewPoint featured functionalities such as digital job sheets, plant tonnage tracking, live job reviews, media attachments, and digital signature capture.

Although the intellectual property for BrewPoint remains with Superior Paving Corp., my role in its end-to-end development demonstrates my capability to solve domain-specific operational challenges through innovative technology. This project is a prime example of how I bring applied innovation to legacy industries such as asphalt and road construction.

In another initiative at Superior Paving Corp., I independently designed and implemented Auto-ShopKeeper — a computerized maintenance management system (CMMS) for managing construction equipment fleets. The system included modules for preventive maintenance scheduling, repair tracking, parts inventory, and customizable audit workflows with photo logging capabilities.

Deployed company-wide, Auto-ShopKeeper significantly reduced equipment downtime, improved service compliance, and centralized maintenance reporting. As with BrewPoint and Auto-ShopKeeper, I do not claim ownership of the software, but I was solely responsible for their development, deployment, and enhancements — underscoring my innovation capability within the constraints of employment-based development.

Similarly, Tim Sylvester, the founder and CEO of Integrated Roadways, emphasizes the benefits of integrating charging coils into concrete pavement, which can be used in both parking lots and on-street parking. This groundbreaking "Smart Pavement" utilizes precast concrete panels that facilitate wireless vehicle charging while also incorporating digital technology and fiber optic connectivity for real-time traffic monitoring. Sylvester mentions that typically, only one out of every five to eight panels needs charging coils to effectively power moving vehicles. The modular design of the pavement allows for easy upgrades and includes advanced features such as Wi-Fi, 5G, and cloud services. The construction process is enhanced, requiring fewer workers and simplifying repairs; damaged panels can be replaced with ease. Demonstration projects in Colorado, Kansas, and Missouri have successfully tested the technology and construction techniques, setting the stage for larger implementations that highlight these advantages.

Asphalt with embedded wireless charging involves incorporating coils into the road surface that wirelessly transmit energy to electric vehicles equipped with a compatible receiver. This technology, also known as inductive charging or wireless power transfer, allows EVs to recharge while driving on designated roads, eliminating the need for plugging in.

The automotive industry is currently exploring two applications for this technology: a wireless charging infrastructure integrated into roadways and pad-based wireless charging systems. The "Arena of the Future," a 1,050-meter-long circuit powered by 1 MW of electricity, has successfully completed its construction and technical implementation. It is now ready for EV field testing to assess the feasibility of this innovative technology.

A skilled team is utilizing the Dynamic Wireless Power Transfer (DWPT) system, which appears to effectively address the ongoing needs for decarbonization and environmental sustainability within the mobility sector. Inductive charging technology represents a viable solution toward achieving the objectives outlined in the publication "Decarbonizing Transportation, Scientific Evidence and Policy Proposals," recently presented by experts from MIMS (Ministry of

Sustainable Infrastructure and Mobility) as part of the "Structure for the Green Transition of Mobility and Infrastructure" (STEMI). This effort is aligned with the discussions in the European Council regarding the Fit for 55 Package, introduced by the European Commission in support of the Green Deal strategy.

Wireless charging technology offers several benefits for electric vehicles, including dynamic charging that allows for continuous power as the vehicle moves, alleviating range anxiety while enhancing efficiency. It can reduce charging times compared to traditional methods, particularly beneficial for high-traffic areas and public transportation. This technology enables the development of electrified highways, potentially transforming transportation into a more sustainable model while lowering infrastructure costs by reducing the need for extensive charging stations.

Notable developments include Electreon's systems in Detroit, research by Purdue University funded by INDOT and the National Science Foundation, and Norway's integration of Wireless charging technology offers several benefits for electric vehicles, including dynamic charging that allows for continuous power as the vehicle moves, alleviating range anxiety while enhancing efficiency. It can reduce charging times compared to traditional methods, particularly beneficial for high-traffic areas and public transportation. This technology enables the development of electrified highways, potentially transforming transportation into a more sustainable model while lowering infrastructure costs by reducing the need for extensive charging stations.

There are numerous benefits and possible applications of EV inductive charging. The reduction of dwell times associated with vehicle charging is set to significantly enhance the travel experience for passengers. This innovative system facilitates seamless integration of a variety of charging technologies, which means that travellers can spend less time waiting for their vehicles to charge and more time enjoying their journeys. The efficiency brought about by this advancement is particularly beneficial in our fast-paced world, where convenience and quick turnaround are highly valued. As a result, this new charging infrastructure stands to elevate the travel standards, making it more appealing and practical for individuals on the go.

Moreover, the technology benefits greatly from the advancements provided by cutting-edge 5G connectivity and artificial intelligence. These technologies enable real-time communication and data exchange between vehicles and management systems, optimizing operations and improving overall efficiency. This system's dynamic and static induction capabilities enhance its versatility, allowing it to be implemented not only on traditional roadways and highways but also across various infrastructures, including ports, airports, and parking facilities. The growing interest in these applications suggests promising prospects for commercial developments, both within Italy and internationally, paving the way for a more interconnected and efficient transportation future.

France aims to electrify 9,000 kilometers of roads by 2035 as part of its strategy to decrease emissions from road freight transport. This initiative will utilize Electric Road Systems (ERS) that enable electric vehicles to charge while in motion. The primary objective is to achieve net-zero emissions in transportation, with an emphasis on heavy-duty trucks and long-haul logistics. ERS technology has the potential to lower CO2 emissions from road freight transport by 86% compared to diesel. Additionally, France is making investments to enhance electric mobility, including the expansion of public charging infrastructure. .

However, challenges such as high implementation costs, ensuring coil durability and maintenance, addressing safety concerns like electromagnetic interference, and achieving compatibility across all EVs need to be addressed for successful widespread adoption.

Inductive charging is not a new concept; it has been around for years in various pilot projects and specific applications. A notable field where this technology has been utilized is in driverless transport systems, particularly in factory settings and warehouses. Additionally, cities like Turin and Genoa in Italy have employed inductive charging for public electric buses since 2002. These buses can recharge quickly at bus stops through charging devices integrated into the road, allowing them to complete their daily routes without needing extensive battery capacities or additional trips to charging stations.

This system significantly reduces the required battery size and weight for the buses, resulting in lighter vehicles and enhanced economic efficiency. As a result, many cities worldwide are now exploring similar projects that leverage inductive charging technology for public transportation. In Germany, for example, initiatives in cities like Augsburg, Braunschweig, and Mannheim are currently underway, with further expansions planned in Berlin since the summer of 2015.

Research is actively underway to develop asphalt that supports wireless charging for electric vehicles, focusing on improving the efficiency and reliability of inductive power transfer. This involves exploring materials and designs that can guide magnetic flux more effectively, reducing energy loss and enabling dynamic charging while the vehicle is in motion.

Inductive Power Transfer (IPT) is a key technology being explored for wireless charging on roads. It utilizes coils embedded in the pavement to create an electromagnetic field, which is then picked up by a receiver on the vehicle, enabling energy transfer without wires. Current research aims to enhance the efficiency of this process by optimizing the design of both the pavement and the charging coils, as well as the materials used in their construction.

One innovative approach involves designing a "pathway" for magnetic flux within the pavement, allowing for more effective energy transfer between the charging coils and the vehicle. This pathway is created using materials with specific magnetic properties, which are strategically placed to maximize efficiency and minimize energy loss during the charging process.

The exploration of various materials, such as ferrites and iron powder, is crucial for improving IPT efficiency. Additionally, the structural design of the pavement—including the depth and spacing of the coils—is vital. Researchers employ finite element modeling to simulate electromagnetic fields and heat dissipation within the pavement, ensuring optimal performance and preventing overheating.

Several challenges remain in the implementation of IPT technology. Key issues include maintaining the mechanical strength of the pavement to withstand traffic loads and effectively managing the heat generated by the charging coils to prevent overheating. Researchers are continuously working to enhance the efficiency of IPT systems, aiming to reduce energy losses. Furthermore, developing cost-effective and scalable solutions is essential for the widespread adoption of wireless charging on roads.

While the concept of charging private electric vehicles at stop lights is theoretically possible, especially in urban areas, several technological and standardization challenges must first be addressed. Effective communication between the vehicle and charging infrastructure is crucial for ensuring accurate energy tracking and legal billing, as well as preventing fraud and misuse. These requirements are not only vital for inductive charging but also apply to traditional plug-in charging stations. via cable at charging stations.

Several prominent research projects are underway. The Rutgers Center for Advanced Infrastructure and Transportation (CAIT) is investigating electrified pavements for wireless EV charging, while the Indiana Department of Transportation (INDOT) is collaborating with Purdue University on a pilot project to create a quarter-mile stretch of road with embedded charging. Additionally, Oak Ridge National Laboratory is developing high-speed wireless chargers for electric vehicles. Overall, this research aims to create a sustainable and efficient charging infrastructure for EVs, enabling a future where vehicles can charge on the go without needing to plug in.

Countries like Sweden, Germany, France, and the United States are actively researching and developing asphalt-based systems for the wireless charging of electric vehicles (EVs).

Sweden is at the forefront with plans to construct the world's first permanent electrified road, connecting logistic hubs between Örebro and Hallsberg, with completion expected by 2025.

Germany is looking into the electrification of sections of the Autobahn, potentially through overhead cables or inductive charging infrastructure, recommending the installation of thousands of kilometers of such systems.

France aims to electrify thousands of kilometers of roadways by 2035, as mentioned earlier, using a combination of overhead cables, rails, and induction charging.

In the United States, research is expanding in regions like Utah, where high-power stationary wireless charging technology for heavy-duty vehicles is being tested, alongside Michigan's plans for a wireless in-road charging system. A notable project involves Electreon collaborating with Utah State University for research funded by the National Science Foundation.

Other Countries include New Zealand, which is pioneering electrified roads through its Inductive Power Transfer (IPT) roadway project, and South Korea, with the KAIST OLEV project focused on developing and bringing wireless charging systems to market.

Research is being conducted on developing concrete that will allow optimal charging for maximum commercial benefits. Magnetizable concrete is an innovative composite material that combines magnetic particles, such as ferrite or magnetite, with a cement matrix, resulting in enhanced magnetic properties. This unique blend offers exciting applications, especially in wireless power transfer (WPT) systems, where its high magnetic permeability makes it ideal for channeling magnetic fields. By incorporating magnetizable concrete into road infrastructure, dynamic wireless charging for electric vehicles becomes feasible, allowing vehicles to charge while in motion and reducing the need for stationary charging stations. Moreover, this versatile material can also be utilized in construction for creating magnetic foundations, as well as in industrial applications and environmental remediation.

The electric vehicle (EV) charging industry plays a crucial role in the transition towards cleaner, more sustainable transportation solutions. As electric vehicle adoption surges worldwide, the demand for a reliable and expansive charging infrastructure continues to grow. However, building and scaling this infrastructure presents unique challenges that industry leaders must confront head-on. Leading in this dynamic environment requires specific skills and strategies that can significantly influence the future of clean transportation.

First and foremost, driving innovation is essential in the EV charging sector. The rapid pace of technological advancement necessitates that leaders stay ahead of the curve, fostering a culture that encourages creativity and experimentation. This means embracing new technologies, such as fast-charging systems, smart grid integration, and user-friendly mobile applications for locating and accessing charging stations. By promoting innovative solutions, companies can enhance the customer experience while also improving the efficiency and sustainability of their operations.

Navigating regulatory complexity is another critical challenge that leaders face. As governments around the world implement various policies and incentives to promote electric vehicle usage, the regulatory landscape becomes increasingly intricate. Leaders in the EV charging industry must develop expertise in local and national regulations, ensuring compliance while advocating for favorable policies. Understanding the nuances of incentives for EV charging infrastructure can provide competitive advantages, allowing companies to capitalize on available resources and support.

Attracting and retaining top talent is paramount for any organization, but it is particularly vital in the rapidly evolving field of

EV charging. The industry requires a diverse skill set ranging from engineering and software development to marketing and customer service. To build effective teams, leaders must create an environment that not only fosters professional growth but also aligns with the values and aspirations of employees. Offering training opportunities, competitive compensation, and a clear career path can help retain the talent necessary for innovation and success.

Developing a long-term strategic vision is essential for keeping pace with the evolving landscape of clean transportation. Leaders must have the foresight to anticipate market trends and adapt accordingly. This vision should encompass not just immediate goals and milestones but also a broader understanding of the industry's trajectory. Collaborating with various stakeholders, including governments, businesses, and nonprofits, can help identify emerging opportunities and establish sustainable practices that benefit the entire ecosystem.

Building partnerships for sustainable growth is another cornerstone of effective leadership in the EV charging sector. This industry thrives on collaboration, whether it's partnering with vehicle manufacturers, energy providers, or technology innovators. Strong partnerships can facilitate knowledge sharing, resource pooling, and strategic alignment, creating a unified approach to expanding charging infrastructure. By working together, stakeholders can accelerate the development and deployment of charging solutions that meet consumer needs and promote a greener future.

In conclusion, successful leadership in the electric vehicle charging industry hinges on addressing the multifaceted challenges of a rapidly evolving market. By driving innovation, navigating regulatory frameworks, attracting and retaining talent, developing a strategic vision, and building strong partnerships, leaders can shape the future of clean transportation. As the world continues to shift towards sustainable mobility, the importance of effective leadership in this sector cannot be overstated. Embracing these leadership essentials will be crucial in overcoming obstacles and realizing the full potential of electric vehicles, ultimately fostering a cleaner, more sustainable future for all.

CHAPTER 11

ADDRESSING INDUSTRY CHALLENGES

> **"Burn 212" is *the concept of "burning" through challenges at 212°F that reminds us to go the extra mile in everything we do. Whether in our careers, relationships, or community involvement, giving that little bit more can lead to significant positive changes. This dedication to excellence, striving for perfection, and continual effort can transform our experiences and results.***
>
> ***Frank Surface,***
>
> ***Chairman, Superior Paving Corp.***

The future of the asphalt workforce is in attracting and training the next generation of paving professionals. The construction industry is facing increasing challenges in manning their crews. Data from 2019, as reported by NAPA, indicates that 91 percent of construction companies were looking for workers. Eighty to 94 percent of those companies also reported that workers were hard to find.

Asphalt paving plays a crucial role in modern infrastructure, yet it faces several persistent challenges, including material degradation, environmental impact, and rising maintenance costs. These issues can compromise road quality, safety, and longevity. However, emerging technologies such as smart sensors, advanced materials, and automated paving systems are transforming the industry. This introduction explores the key challenges in asphalt paving and highlights how innovative solutions are paving the way toward more efficient, durable, and sustainable road construction.

One major challenge is inadequate funding. Many projects are halted or delayed due to budget constraints, which can lead to increased costs over time. Governments and private entities must work collaboratively to secure sustainable funding sources that ensure timely project completion.

Another significant challenge is the skilled labour shortage, as mentioned above. As experienced workers retire, the influx of new talent into the industry is not enough to fill the gap. This shortage can lead to project delays and compromised quality. Investing in training programs that equip new workers with the necessary skills is crucial to combating this issue. Partnerships between construction companies and educational institutions can facilitate a well-prepared workforce to meet industry demands.

Moreover, the industry must navigate environmental regulations and sustainability concerns. As society becomes increasingly aware of its carbon footprint, road construction projects are expected to adopt more sustainable practices. This includes utilizing eco-friendly materials, minimizing waste, and implementing efficient construction processes. Adapting to these regulations can increase initial costs and complicate project timelines. However, by embracing innovative technologies and sustainable practices, the industry can enhance its long-term viability and public image.

Lastly, the integration of technology poses both opportunities and challenges. Advances such as Building Information Modeling (BIM) and automated machinery can significantly enhance efficiency and reduce errors. However, the initial investment in these technologies can be steep, and the workforce must be adept at using them to realize their full potential.

Addressing these challenges requires collaboration among stakeholders, investment in workforce development, commitment to sustainability, and the integration of innovative technologies. By proactively tackling these issues, the road construction industry can continue to evolve and contribute to modern infrastructure development.

The workforce shortage and the need for skilled professionals to operate and manage the increasingly technological aspects of asphalt paving are significant hurdles the industry must overcome. The adoption and effective implementation of the latest technologies in asphalt paving are heavily reliant on a skilled and knowledgeable workforce. Without attracting and properly training the next generation, the industry risks hindering its progress and the successful deployment of these advancements.

A significant portion of the current asphalt paving workforce is nearing retirement age, leading to a loss of valuable experience and expertise. This poses challenges as it leads to the loss of valuable experience and expertise that younger workers may not yet possess. Organizations need to address these challenges by planning for succession, retaining experienced workers, and adapting to changing workforce demographics.

Older workers often have extensive practical experience and knowledge that younger workers are still acquiring. This can lead to a gap in skills and understanding when they retire, potentially impacting project quality and efficiency. The need to prepare for and train younger workers to take over the responsibilities of retiring professionals is crucial. This requires proactive efforts to mentor and develop the next generation of employees. Older workers may also be more prone to health issues affecting their ability to work. Additionally, they may be less inclined to adapt to new technologies and processes. Older workers may face challenges related to physical stamina, mobility, and the ability to perform physically demanding tasks, and find it difficult to transition to new roles or technologies.

Another persistent challenge for the construction industry, including asphalt paving, is facing negative perceptions due to perceived physical demands, outdoor work in varied conditions, and a lack of perceived technological sophistication. These perceptions can make the industry seem less appealing to potential workers and may lead to a shortage of skilled labor.

Construction work, particularly asphalt paving, requires significant physical exertion, including lifting heavy materials, working in awkward positions, and spending long hours outdoors. Construction projects are often exposed to harsh weather conditions like extreme heat, cold, rain, and wind. This can deter some individuals, especially those seeking careers with less physical strain. While the construction industry is increasingly incorporating technology, some may still perceive it as a traditional industry relying on manual labor and outdated methods, making it less appealing to those drawn to technology-driven fields. The negative perceptions can create a barrier for the industry to be seen as a modern and innovative sector, hindering efforts to attract investment and improve productivity.

In recent years, the asphalt industry has experienced a significant transformation driven by technological advancements. However, younger generations often lack awareness of these changes, which can lead to a neglect of potential career paths within this sector. Many youths may have preconceived notions about the asphalt industry, viewing it as outdated or labor-intensive without recognizing the modern capabilities that technology has introduced. Innovations such as Building Information Modeling (BIM), automation, and advanced data-driven methodologies have revolutionized how projects are managed and executed. These developments enhance efficiency and accuracy and improve safety standards on job sites. As younger individuals explore career options, it is crucial to bridge the gap between traditional perceptions and the reality of an industry that is evolving with technological integration.

Traditional training programs often fall short in addressing these advanced requirements, leaving many workers unprepared for the modern demands of the industry. For instance, proficiency in AI-powered software, familiarity with advanced sensors, and an understanding of automated systems are increasingly essential for those entering the workforce. To attract and retain talent, the industry must invest in education and training initiatives aligning with current technological trends. Building partnerships with educational institutions and establishing targeted apprenticeship programs could help to develop the necessary skills among the workforce, ensuring that the asphalt industry not only adapts to changes but thrives in an increasingly tech-driven environment.

The construction industry is currently facing significant challenges in attracting and retaining talent in a competitive labor market increasingly dominated by other sectors such as technology and engineering. As these fields continue to evolve, they often offer lucrative salaries, flexible work environments, and appealing career advancement opportunities that attract skilled professionals. These factors make it difficult for the construction industry to compete, particularly when the perception of construction jobs can sometimes be less appealing due to physical demands, lengthy hours, and perceived lack of innovation. This competition causes a talent drain, where engineers, project managers, and skilled tradespeople gravitate toward industries that offer more modern work practices and advanced technological integration. In essence, the construction sector must not only vie for individuals already considering the field but also think critically about how to present its unique offerings in a way that resonates with a new generation that values workplace culture and technological engagement.

Even when the construction industry successfully attracts talent, the challenge doesn't end there; retaining that talent is equally, if not more, important. Competitive compensation is crucial, but it's not the sole factor that influences employee retention. To keep staff engaged and motivated, companies need to provide clear opportunities for personal and professional growth. This can be achieved through training programs, mentorship, and pathways to advancement that empower employees to envision a long-term career within the industry. Furthermore, fostering a positive work environment that embraces technology, collaboration, and innovation is essential. Firmly integrating modern tools and practices can create a more satisfying and efficient workplace, making roles in construction more appealing. Emphasizing a culture of safety, respect, and support also leads to increased job satisfaction and loyalty among workers.

The industry must adapt to these evolving expectations, becoming more user-friendly and tech-savvy to ensure that talent not only joins but stays committed to building the future through construction.

To attract workers to our industry, negative perceptions of the industry need to be rebutted. Pavement construction is generally seen as a dirty and dangerous work environment. Targeted messages about the skills needed, industry opportunities, pay equity and diversity of our workforce have been shown to improve perceptions of our great industry by double digits. Concerted efforts with educators are extremely valuable in improving perceptions and attracting the next generation to construction.

Efforts to recruit future engineers and scientists to the asphalt world must be encouraged. Organizations such as the Asphalt Institute Foundation and its education programs can pay untold dividends in developing these needed resources.

America has implemented various programs and initiatives to train the next generation of road paving workers, including focusing on technical training centers, promoting career opportunities, and utilizing advanced technologies. Wirtgen America's Training & Technology Center is a key example, connecting the road construction industry with technical training and professional development.

Institutions like Wirtgen America's Training & Technology Center offer technical training and professional development to owners, operators, dealers, and service technicians, focusing on perfecting skills and promoting a passion for road building. These centers also connect the industry with career opportunities, promoting the road construction profession. The industry is embracing advancements in road safety, using robotics, geospatial mapping, and AI to enhance efficiency and safety. There's a growing emphasis on environmentally sustainable practices, including efficient drive trains that reduce fuel consumption. These centers facilitate connections within the road construction industry, fostering collaboration and knowledge sharing. Research and development efforts are focused on creating innovative pavement materials, including those with energy-harvesting capabilities or enhanced durability.

There are some tech-driven solutions for the challenges discussed so far. The road construction industry is rapidly evolving, with technology playing a pivotal role in enhancing efficiency, safety, and training methodologies. As the demand for skilled labor increases and the complexity of projects escalates, tech-driven solutions are becoming indispensable. Technology trends are reshaping road construction, focusing on Virtual and Augmented Reality (VR/AR) for training, Building Information Modeling (BIM), online learning platforms, gamification, data analytics, collaborative tools, and marketing strategies to attract a younger workforce.

One of the most innovative approaches to training in road construction is the incorporation of Virtual and Augmented Reality (VR/AR). These immersive technologies allow for the simulation of real-world scenarios, preparing workers for the challenges they will face on-site without the inherent risks.

VR can create a realistic environment where trainees can experience pavement scenarios and the operation of machinery such as pavers and rollers. Meanwhile, AR can enhance on-the-job training by overlaying digital information onto physical equipment, guiding workers through maintenance and troubleshooting tasks. Various companies are developing VR training modules that enable new hires to practice operating machinery in a simulated environment before they step onto a live construction site. Furthermore, AR applications are emerging to assist workers in understanding complex machinery by providing real-time maintenance procedures and guidance.

Building Information Modeling (BIM) and digital twins are revolutionizing how education in road construction is delivered. These technologies facilitate the visualization of projects, allowing students to comprehend the construction process from inception to maintenance. BIM models enable learners to interact with detailed project data, fostering a comprehensive understanding of workflows in construction. Digital twins, on the other hand, provide real-time data that can be analyzed for informed decision-making, requiring a new age of professionals skilled in interpreting this information. Educational institutions have begun integrating BIM software into their civil engineering and construction management curricula. This approach not only equips students with the necessary skills but also familiarizes them with the digital workflows that dominate the modern construction landscape.

Accessibility and flexibility in education are crucial, particularly in an industry that demands continual learning. Online learning platforms make it easier for professionals to enhance their skills from anywhere. Online training modules cover a wide array of topics—including new technologies, software, and best practices—enabling workers to stay competitive in the job market. Furthermore, remote expert support can be delivered through video conferencing and AR, enhancing the learning experience. Industry associations, alongside companies, are launching online courses focusing on crucial aspects such as drone surveying, data analysis for pavement management, and the operation of automated construction equipment. These courses are designed for learners' needs while allowing flexible schedules.

With younger generations increasingly accustomed to interactive digital experiences, gamification offers a fresh approach to training in road construction. By incorporating game-like elements into training, organizations can increase engagement and retention among trainees, making the learning experience enjoyable. This approach can be particularly effective in teaching complex tasks such as optimizing paving patterns or coordinating material flow on a construction site. Various simulation games are being developed to help workers practice and refine their skills in a virtual setting. Such innovations boost motivation and enhance skill acquisition in a fun and interactive way.

Data analytics and AI are transforming training program deliveries, offering personalized learning experiences that cater to individual skill levels and progress. By tracking learners' progress, AI-powered platforms can identify specific skill gaps and tailor training programs accordingly. This personalized approach allows workers to focus on areas needing the most support, accelerating their learning journey. Learning management systems that utilize AI can adapt content based on a user's performance, delivering recommendations for relevant materials that align with their learning needs.

Effective communication and collaboration are essential in any construction project, and digital tools facilitate this process. By utilizing communication platforms that support knowledge sharing, road construction teams can foster a more connected and collaborative work environment, which is particularly appealing to tech-savvy individuals. Project management software integrated with communication features enables teams to share data, discuss issues, and learn from each other in real-time, enhancing overall productivity and teamwork.

A younger workforce can be attracted by utilizing digital media for marketing and outreach. It is one of the most effective strategies to showcase the technological advancements in the field. Industry stakeholders can harness social media, online videos, and other digital channels to highlight technological innovations and career opportunities, thereby changing perceptions of the asphalt industry. Campaigns spotlighting the use of drones, robotics, and data analytics in asphalt paving have proved successful in attracting new talent, promoting a modern image of the industry.

The focus is clear: to equip the next generation of construction professionals with the skills they need to thrive. By embracing these tech-driven solutions, the industry can enhance training, improve workforce competency, and attract a new wave of talent eager to make their mark in road construction.

The future of the asphalt workforce is bright for another reason. The asphalt industry workforce of the future will likely be characterized by a greater emphasis on technology, automation, and specialized skills, while still retaining the need for human expertise in certain areas. This includes using AI and robotics to automate repetitive tasks, improve efficiency, and enhance safety. Simultaneously, the workforce must adapt to new materials, construction techniques, and environmental initiatives.

AI-powered tools and robots can automate tasks like asphalt mix design, paving, and quality control, freeing up human workers for higher-level tasks and strategic decision-making. The workforce will need to develop skills in software development, robotics maintenance, data analysis, and using new materials and construction techniques. The industry will continue to embrace green initiatives, including the use of recycled materials (RAP), carbon reduction strategies, and innovative asphalt binders, requiring workers to be knowledgeable about these practices. Organizations need to prepare their workforce for a dynamic future by fostering adaptability, encouraging re-skilling, and providing opportunities for continuous learning. New technologies and processes must be used to enhance safety and reduce physical strain on workers, particularly those involved in manual tasks. While technology will automate some tasks, human expertise will still be crucial for tasks requiring judgment, problem-solving, and complex decision-making, such as overseeing construction projects and troubleshooting issues. Companies must prioritize employee well-being and engagement as they adapt to technological changes, ensuring that employees feel supported and valued.

In essence, the asphalt industry workforce of the future will be a blend of skilled professionals, technicians, and specialists who can effectively utilize technology to achieve efficiency, sustainability, and safety while retaining the human touch needed for complex tasks and problem-solving.

Another challenging aspect is the sustainability imperative. The challenge of a high carbon footprint in asphalt production arises from energy-intensive processes such as heating, binder production, and transportation, which significantly contribute to greenhouse gas emissions, leading to increasing regulatory pressures and public scrutiny. Solutions include the utilization of increased recycled asphalt pavement (RAP) and recycled asphalt shingles (RAS) optimized through AI-driven mix design software, and real-time quality control sensors for consistency. Additionally, warm mix asphalt (WMA) technologies utilize smart temperature control for energy efficiency, and digital twins of asphalt plants facilitate energy optimization. Exploring alternative binders from bio-based materials and low embodied carbon aggregates, supported by AI in material science, further enhances sustainability efforts. Furthermore, the electrification of construction equipment, with smart charging infrastructure, and carbon capture and utilization (CCU) technologies to repurpose CO2 emissions into usable materials, represents innovative strategies to reduce the overall environmental impact of asphalt production.

Resource depletion and waste generation in the construction industry, particularly the finite reserves of high-quality aggregates and the increasing volume of construction and demolition waste, such as old asphalt pavements and shingles, that end up in landfills, are huge challenges. This situation is exacerbated by the rising costs and scarcity of suitable aggregates in some regions, alongside limited landfill space. To tackle these challenges, several tech-driven solutions have emerged, including advanced material characterization utilizing AI and machine learning for better analysis and utilization of recycled asphalt pavements (RAP) and asphalt shingles (RAS). Robotic sorting and processing in recycling plants enhance the efficiency of sorting and processing construction waste for use as recycled aggregates, while circular economy platforms enable the connection of waste material sources with potential users in the construction industry, promoting material reuse. Additionally, implementing Building Information Modeling (BIM) and digital material passports facilitates tracking of pavement layer compositions, making recycling and reuse at the end of a material's life easier and more efficient.

Variability in materials and weather can lead to inconsistent pavement quality, resulting in issues like potholes and cracking. To tackle these, tech-driven solutions are being deployed, including real-time quality control via integrated sensors on paving equipment, and AI-powered image analysis for defect detection. Non-destructive testing methods like ground-penetrating radar assess pavement conditions, while blockchain ensures transparency in material tracking. Predictive performance modeling uses AI to analyze data, optimizing mix designs and maintenance schedules.

The growing frequency of extreme weather poses significant challenges for asphalt pavements, necessitating heightened maintenance and reducing lifespan. Innovative solutions include climate-resilient pavement designs using AI-optimized materials and permeable pavement technologies for better stormwater management. Integrating advanced weather monitoring and smart drainage systems can adapt to changing rainfall patterns. Addressing the urban heat island effect, cool pavement technologies enhance surface reflectivity, while green infrastructure and drone-based thermal mapping help identify heat islands and promote effective cooling strategies.

The increasing frequency and intensity of extreme weather events are a major threat to the longevity and performance of asphalt pavements. Designing for resilience is no longer a niche consideration but a fundamental requirement for future infrastructure.

Recent climate changes have led to an increase in pothole formation due to more frequent freeze-thaw cycles, as well as rutting and softening of roads during prolonged high temperatures. Intense rainfall and flooding are causing washouts and structural damage, while rising sea levels and storm surges are impacting coastal infrastructure. To address these challenges, innovative solutions include AI-optimized mix designs for improved temperature and moisture resistance, advanced materials like polymers and fibers for enhanced pavement strength, permeable pavement technologies for better drainage, smart drainage systems, cool pavement technologies to reduce heat absorption, and advanced weather monitoring with predictive analytics for proactive maintenance.

Moving beyond the initial cost and focusing on the total cost of ownership over the pavement's lifespan is crucial for making informed decisions that benefit both the environment and the economy. It also touches upon the "Ensuring Consistent Quality and

Performance" challenge, as longer-lasting pavements reduce the frequency of costly repairs and replacements.

Budget constraints often lead agencies to choose low upfront cost projects that can result in higher long-term expenses due to increased maintenance needs. As repair and rebuilding costs rise, the economic impacts of road closures and delays become significant, alongside hidden environmental costs from more frequent construction. Utilizing tools like Life Cycle Assessment (LCA) and predictive performance modeling can help in accurately calculating the long-term costs and optimizing maintenance schedules. Additionally, incorporating Building Information Modeling (BIM) enhances design and material selection for better durability. Advanced material characterization and value engineering methodologies further ensure the use of high-quality materials and cost-effective solutions that prioritize long-term performance.

Pavement and maintenance management systems and quality assurance systems are key to providing quality pavements. Pavement and maintenance management systems are mature and useful to those involved in scheduling maintenance operations in the short term and tracking the condition of a public agency's pavements. The use of the information contained in these systems by decision-makers and planners for allocating financial resources for various budgeting horizons needs improvement. Integration of these "big data" systems with construction, materials, environmental and other public agency data systems is needed.

Rapid pavement evaluation equipment capable of operating at highway speeds and collecting continuous data needs further refinements. Advances in devices that use electromagnetic waves to determine pavement thickness, condition, surface texture, and drainage (ground penetrating radar and LiDAR for example) are needed.

Rapid quality control and quality assurance test methods are needed. The industry has largely been unsuccessful in developing meaningful, automated, rapid test methods that can be utilized for equipment production process control and construction quality control. Hand-held instruments that evaluate and classify RAP and provide recommendations for the type and quantity of rejuvenator would be particularly useful.

Improvements in plants and equipment are constantly being undertaken by industry. Asphalt mixture plant production automation, intelligent compaction, automated machine guidance, GPS-based profile milling and laydown of pavement layers, e-ticketing and automated data collection from plant and field records for quality assurance purposes are examples of a few advancements that are underway.

Automated process control and quality control testing equipment that will sample and test materials rapidly will improve the quality of construction. Automated aggregate sampling and gradation testing equipment is an example of needed equipment that has been developed but has been difficult to deploy in the industry. More comprehensive testing and evaluation of in-place materials through the deployment of current and future technology, such as dielectric profiling systems, will provide engineers with more thorough information. Additional developmental efforts are needed in this and other equipment.

Accelerated construction techniques need to be developed. The goal of accelerated construction is to reduce construction project time by as much as 70 percent. This is a large effort involving a workforce skilled in several disciplines, including contracting methods, traffic management, work zone safety, materials, construction operations, economics, and construction equipment. Challenging long-held beliefs of acceptable maximum lift thickness, for example, can speed the delivery of construction projects while potentially improving their performance.

In today's rapidly evolving technological landscape, collaboration between industry stakeholders, researchers, and policymakers is more crucial than ever. The integration of innovative technological solutions can significantly enhance operational efficiencies, drive economic growth, and address pressing societal challenges. However, the successful adoption of these solutions requires a united effort from all involved parties.

One prominent example is the advancement of renewable energy technologies. Companies in the energy sector, such as Tesla and Siemens, have made significant strides in solar power and battery storage. However, achieving widespread adoption necessitates collaboration with researchers who can innovate on the technology's efficiency and storage capacity. For instance, university research teams working on new photovoltaic materials or energy storage solutions can provide invaluable insights that enhance product performance. Simultaneously, policymakers play a critical role by establishing regulations that incentivize the use of renewable energy and create a more favorable market environment. This multifaceted collaboration has led to significant breakthroughs, such as the development of solar panels that are not only more efficient but also cheaper to produce.

Additionally, the field of artificial intelligence (AI) demonstrates the importance of collaborative efforts. Companies like Google and Microsoft are leading the way in AI development, yet to ensure ethical and safe applications of these technologies, they must work in tandem with researchers and regulators. Academic institutions are critical in studying the implications of AI, including biases in algorithms and the socio-economic impacts of automation on the workforce. Policymakers can then create frameworks that govern AI deployment, ensuring that technology benefits society while minimizing harm. For instance, the European Union's AI Act is an example of a regulatory approach to ensure responsible AI development. This act represents a collaborative effort to establish guidelines that protect citizens' rights and promote innovation at the same time.

Furthermore, addressing climate change—a challenge that affects all sectors—requires a concerted effort among stakeholders. Industries are adopting technological solutions like carbon capture and storage, but without the collaboration of researchers who can innovate these technologies and policymakers who can enforce environmental regulations, progress may stall. The effectiveness of these solutions is contingent on research that improves their efficiency and on legislation that prioritizes sustainable practices.

As we move forward in this technology-driven age, fostering such partnerships will be paramount to addressing complex challenges and unlocking the full potential of technological advancements for the benefit of society.

CHAPTER 12

THE STARTUP OPPORTUNITY IN ASPHALT TECH

Asphalt plays a crucial role in our daily lives, whether inside our houses or when we step out, often going unnoticed despite its significant contributions to modern infrastructure and transportation. It is primarily used in road construction, providing smooth surfaces that enhance the safety and efficiency of travel. The durability and flexibility of asphalt make it ideal for withstanding various weather conditions, reducing wear and tear on vehicles. This facilitates smoother commutes and minimizes vehicle maintenance costs, making transportation more accessible and economical.

Beyond roadways, asphalt's relevance extends to various applications in urban planning and development. It is used in parking lots, driveways, and airport runways, showcasing its versatility in supporting different aspects of transportation networks. Asphalt is an integral component of our everyday lives, providing essential support for transportation and urban development. As we navigate through our cities, the asphalt beneath our feet and wheels plays a vital role, facilitating the rhythm of everyday life.

As discussed in the previous chapters, technology plays a crucial role in modern road construction, offering advancements in design, planning, material usage, and execution, leading to faster, more efficient, and sustainable projects. Key technologies include Building Information Modeling (BIM), prefabrication, drones, 3D printing, and automation, all contributing to improved accuracy, cost savings, and project timelines. Software integrates timelines, budgets, and resource allocation into a single platform, facilitating real-time updates and improved communication among stakeholders. Off-site manufacturing of road components can significantly shorten construction schedules, improve quality control, and reduce disruptions. Bulldozers, excavators, and pavers equipped with automation can perform tasks with greater precision and efficiency. These systems combine GPS with digital models to automate earthworks, potentially increasing efficiency by up to 30%. Drones and lidar technology are used for surveying, mapping, and monitoring construction progress. GNSS (Global Navigation Satellite System) and other geospatial tools provide accurate design data, construction, and maintenance. Techniques like laser scanning and photogrammetry create detailed construction site 3D models. Technologies like V2X (Vehicle-to-Everything) communication, sensor systems, and dynamic warning signs can enhance safety and efficiency on roadways.

In this chapter, I aim to present a thorough and forward-thinking analysis of the promising opportunities available for entrepreneurs. This century offers unparalleled avenues for innovation and skill development, particularly in addressing pressing challenges and promoting sustainable development. The asphalt sector, in particular, represents a significant landscape for pilot projects that could lead to impactful advancements.

There are several promising opportunities for startups in the asphalt technology sector, particularly focusing on sustainability and innovation. Companies can explore the use of recycled materials, such as old asphalt shingles, in production to minimize waste and lessen environmental impact. Innovations in asphalt mix design, including options that allow paving in wet conditions or enhance durability, can meet industry challenges and open up new markets. Additionally, advancements such as energy-generating asphalt with embedded photovoltaic cells and the application of nanotechnology to improve strength and flexibility represent exciting avenues for development. Automation and robotics in paving and maintenance can enhance efficiency while reducing labor costs, while leveraging data analytics and AI can optimize asphalt mix designs and manage operations more effectively.

For any asphalt tech startup, several key considerations must be taken into account for success. Conducting thorough market research to understand the needs of potential customers, including construction firms, municipalities, and road maintenance contractors, is crucial. A strong business plan outlining the startup's vision, strategy, target market, and financial projections is essential. Securing funding from various sources, such as venture capital or government grants, will support initial operations and research. Building strategic partnerships with established industry players and ensuring compliance with relevant regulations are important steps, as is protecting innovative technologies through patents and trademarks.

Road construction and urban development are critical components of modern infrastructure that evolve continuously to accommodate the growing and shifting needs of society. As urban populations increase and technology advances, the demand for efficient transportation systems and well-planned urban environments becomes ever more pressing. This necessitates a commitment to ongoing improvements and innovations in construction practices and urban design.

Governments across America, Europe, and developing countries are increasingly promoting entrepreneurship in asphalt technology, creating a ripe environment for innovative ventures and pilot projects. In the United States, initiatives are being launched to advance sustainable infrastructure, supporting entrepreneurs who focus on developing asphalt materials that enhance longevity and reduce environmental impact. Similarly, in Europe, the Green Deal and various national policies aim to foster sustainable construction practices, encouraging startups to explore cutting-edge asphalt technologies that align with climate goals.

In developing nations, the push towards improved infrastructure presents a unique opportunity for entrepreneurs. Governments are recognizing the need for durable and cost-effective road solutions, prompting collaborations with local entrepreneurs who can leverage traditional techniques alongside modern innovations.

These efforts reflect a global trend where both established economies and emerging markets collaborate to find solutions that enhance road safety and sustainability. With the support of government programs, funding, and access to research and development resources, entrepreneurs in asphalt tech are well-positioned to lead the way in transforming transportation infrastructure, thereby improving economies and quality of life worldwide.

Research and development play pivotal roles in identifying sustainable materials, implementing smart technology, and enhancing safety measures in road construction. Furthermore, the integration of environmental considerations into urban planning is paramount, as cities strive to reduce their carbon footprints and promote green spaces.

As societal needs transform in response to economic, cultural, and environmental factors, the dynamic nature of urban development ensures that there will always be opportunities for advancement. Ultimately, the pursuit of a more efficient and livable urban environment will continue to drive investment and creativity in the fields of construction and development.

REFERENCES

1. https://www.hologram.io/blog/digital-twin-technology/
2. https://digital.adb.org/home/2024/8/6/digital-twins-for-smart-road-construction
3. https://swarajyamag.com/infrastructure/china-completes-worlds-first-fully-unmanned-highway-paving-project#:~:text=In%20a%20display%20of%20engineering,using%20drones%20and%20autonomous%20robots.
4. https://gpcsl.com/2024/10/15/robots-pave-the-way-chinas-unmanned-road-resurfacing-milestone/#:~:text=These%20machines%20are%20capable%20of,quality%20throughout%20the%20resurfacing%20process.&text=Working%20in%20tandem%20with%20the,compacted%20the%20newly%20laid%20asphalt.
5. https://www.cece.eu/news/introducing-infrarob-european-project-seeking-for-automated-autonomous-solutions-for-the-integrity-performance-and-security-of-diverse-infrastructures#:~:text=The%20EU%2Dfunded%20InfraROB%20project,road%20construction%20and%20maintenance%20work.
6. https://buildindigital.com/robot-tech-takes-the-high-road-in-austria-motorway-trial/
7. https://newsroom.strabag.com/en/press-releases/strabag-at-1/2024-10/freeway-9#:~:text=Fully%20autonomous%20asphalt%20paving%20and,partners%20in%20the%20EU%2Dwide

8. https://mobilityintelligence.michelin.com/us/practical-applications-of-artificial-intelligence-and-machine-learning-for-better-road-maintenance/#:~:text=Using%20AI%2C%20Michelin%20Better%20Roads,pavement%20maintenance%20planning%20a%20proactive
9. https://www.azobuild.com/news.aspx?newsID=23723#:~:text=By%20optimizing%20asphalt%20mixtures%2C%20ML,costs%20and%20extend%20road%20lifespan.
10. https://www.petronaftco.com/smart-infrastructure-and-ai/
11. https://leotek.com/predictive-maintenance/
12. https://www.constructionplacements.com/10-innovative-uses-of-green-cement-in-construction/
13. PetroNaft https://www.petronaftco.com › Article
14. The Role of AI in Reducing Labor Shortages in Construction by Tim Safransky, CPA, VP at BZ Construction Management, LLC
15. The Role of AI in Alleviating Labor Shortages in Construction by Tim Safransky, CPA, VP at BZ Construction Management, LLC
16. Case Studies of Implementation- IndiaRAP
17. https://english.mathrubhumi.com/
18. MDPI stands for Multidisciplinary Digital Publishing Institute.
19. Presentation on SlideShare, Waste Materials Used in Road Construction.
20. Yale e360.

21. Life cycle assessment of roads: Exploring research trends and harmonization challenges, ScienceDirect
22. Raposo et al., 2019
23. https://www.acpa.org/magazine/innovations-inductive-charging-in-pavement/
24. https://cdn.vector.com/cms/content/know-how/_technical-articles/EMobility_Inductive_Charging_EmobilityTec_201510_PressArticle_EN.pdf
25. EVBoosters
26. https://evboosters.com/ev-charging-academy/career-hiring-tips/key-essentials-for-leadership-in-ev-charging/
27. https://www.tensarinternational.com/resources/articles/what-causes-potholes#:~:text=Stage%204%20%E2%80%93%20Freeze%2Dthaw%20weather%20worsens%20potholes&This%20leads%20to%20faster%20development%20and%20deeper%20potholes.
28. https://www.epa.gov/heatislands/using-cool-pavements-reduce-heat-islands
29. https://openwage.com/how-to-attract-talent-tips-for-the-construction-sector/#:~:text=This%20is%20especially%20true%20in,where%20employers%20compete%20for%20talent.
30. https://www.asphaltmagazine.com/future-challenges/
31. https://www.transportation.gov/

32. https://www.asphaltpavement.org/
33. https://www.asphaltpavement.org/climate
34. https://www.asphaltpavement.org/expertise/sustainability
35. https://member.asphaltpavement.org/Shop/Product-Catalog/Product-Details?productid={8144A43C-57CA-EC11-A7B5-000D3A1460CF}
36. https://www.vdot.virginia.gov/
37. https://timsylvester.com/media/
38. https://www.energy.gov/sites/default/files/2023-01/the-us-national-blueprint-for-transportation-decarbonization.pdf
39. https://www.epa.gov/
40. https://www.novatr.com/blog/road-construction-innovations
41. https://imaginovation.net/blog/construction-industry-technology-trends/#:~:text=Building%20Information%20Modelling%20(BIM)&text=Another%20hot%20construction%20technology%20trend%20is%20BIM,in%20an%20open%20and%20highly%20collaborative%20environment
42. https://americanasphalt.com/how-is-asphalt-paving-recycled/#:~:text=Supporting%20data%20from%20the%20U.S.,percent%20of%20which%20is%20recycled
43. https://pavementrecyclers.com/what-is-recycled-asphalt/
44. https://www.mdpi.com/2227-9717/11/7/2095#:~:text=The%20application%20of%20rec

ycled%20materials%20in%20asphalt,as%20well%20as%20carbon%20dioxide%20emissions%20[3]

45. https://www.startuphustlepodcast.com/navigating-the-future-tech-in-asphalt-paving/
46. https://www.tensarinternational.com/resources/articles/what-causes-potholes#:~:text=Stage%204%20%E2%80%93%20Freeze%2Dthaw%20weather%20worsens%20potholes&This%20leads%20to%20faster%20development%20and%20deeper%20potholes.
47. https://www.epa.gov/heatislands/using-cool-pavements-reduce-heat-islands

ABOUT THE AUTHOR

Pruthvi Yannam is a tech innovator and researcher whose work bridges artificial intelligence, road safety, and sustainable infrastructure. With a Master's in Information Systems from George Mason University and hands-on experience as a Developer with Superior Paving Corp (USA), his research focuses on predictive maintenance, autonomous pavement systems, and smart safety zones.

He is a builder of smarter roads and greener futures — a curious software engineer/developer at heart with a passion for practical problem-solving. Raised in India and now working in the U.S., he grew up surrounded by the challenges of growing cities and uneven infrastructure. Somewhere between dodging potholes and watching late-night TED Talks, he imagined how technology could solve problems we all quietly accepted.

He is the creator of tools like PaveIQ and EcoQuant, and has authored multiple patent-pending systems addressing urban emissions, carbon credit tracking, and edge-based infrastructure intelligence. Some of his publications and pilot projects, still in progress, for Indian and international circles will contribute hugely to SDG-aligned smart cities.

Through real-world deployment and open-access research, Pruthvi advocates for future-ready infrastructure that is intelligent, equitable, and environmentally conscious.

He is deeply committed to working on innovative technologies and products that address pressing global challenges. His enthusiasm for development drives him to continuously seek knowledge and enhance his skill set. He embraces the principle of lifelong learning and is eager to acquire any new technologies or platforms that can augment services or products effectively.

While he does not claim to possess all knowledge or expertise, he assures of his dedication to learning and willingness to invest significant effort to master any task at hand. He firmly believes that with determination and hard work, one can achieve remarkable outcomes and overcome obstacles that may initially seem insurmountable.

His primary interest lies in developing solutions that simplify processes and contribute to a better quality of life, ultimately creating a more sustainable and less stressful environment for all.

From Asphalt to AI is his first book — a bridge between that moment of wonder on a dusty street corner and the world of ideas he's spent nearly a decade bringing to life.

With a background in information systems and experience in the asphalt and paving industry, Pruthvi is currently designing tools that help cities think ahead — from predicting when a road will crack to tracking the emissions a construction site leaves behind. His projects reflect a simple belief: that better infrastructure means a better life for everyone who walks, drives, or dreams on it.

When he's not building AI models or dashboards, you might find him wandering through new cities, watching their road and traffic patterns, sustainable infrastructure, or writing about the world he wants to help shape.

From Asphalt to AI is his way of bringing that vision into the open — a guidebook for anyone who believes the future should be smoother, smarter, and more sustainable.

Verses Kindler Publication

Verses Kindler Publication

Reach us through our website -
https://www.verseskindlerpublication.com/
For more information visit our
Instagram or Facebook page.